WEATHER AND SUPPLY BEHAVIOUR IN AGRICULTURE

AN ECONOMETRIC APPROACH

by

Dr. Gursharan Singh Kainth

Regency Publications

New Delhi 110 008

ISBN 81-86030-22-0

Published by Regency Publications, 20/36-G, Old Market, West Patel Nagar, New Delhi 110 008 and printed at Radiant Printers, New Delhi, Phones: 571 3968, 578 3571.

सत्यमेव जयते

MEMBER
PLANNING COMMISSION
YOJANA BHAVAN
NEW DELHI 110 001

FOREWORD

The basic objective of Dr. Kainth's empirical study is to examine the crop-weather relationship in Punjab state, the craddle of Green Revolution. It is a matter of great concern as to whether the acceleration in the growth of crop output following the Green Revolution has been tapering off in this agriculturally advanced state of the Indian Union. Punjab agriculture has been operating at high input-output equilibrium as compared to other states. If there is no scope for raising productivity further, Punjab agriculture would become high cost economy deflecting farm investments. It is in this context, Dr. Kainth's observation, that there exist vast untapped potential of wheat output in some of the districts like Ludhiana and Gurdaspur is quite heartening.

Dr. Kainth's work represents a rigorous yet lucid analysis of the influence of non-economic factors on production pattern in Punjab. The study makes interesting observations on the pattern of rainfall in the Punjab state and offers very useful hints for prediction. The study amply demonstrated that the belief that the technological innovations such as assured irrigation can ensure the freedom from the clutches of the weather vagaries is a myth. The study also comes out with a conclusion that supply response to price is positive in the case of main cereals i.e. wheat and paddy.

The author also provides a very comprehensive review of literature on the supply response in agriculture. Thus, we have a very timely and valuable study, carefully conceived and executed on important questions.

I am sure the book will be an invaluable source of reference not only to planners, policy-makers and administrators but to the students of agriculture alike. The general readers who are interested in Indian agriculture in general and Punjab agriculture in particular will find the book very useful.

(Jayant Patil)

PREFACE

Agriculture industry in India is proverbially called 'a gamble on the monsoon'. Full potential of climate as an agricultural resource has not been used or even realised. Apparently several crops are grown traditionally in areas without any consideration for climate suitability. As a result, poor yield of crops are obtained and at the same time much of the production potential of this vast natural resource is left un- or underutilised. The present study is an empirical investigation of the crop weather relationship in one of the agriculturally advanced state of the Indian Union.

In a work of this size and nature, progress is impossible without the genuine help of many researchers and scholars. Discussions with Dr. Rajinder Singh Bawa, Dr. Joginder Singh, Dr. Prem Lal Mehra, Dr. Sukhwant Singh Bhinder and Dr. Meharban Singh were an unfailing source of intellectual stimulation. A number of improvements are due to the candidness with which they made suggestions. My sincere and profuse appreciation is due to Dr. Jayant Patil, Member, Planning Commission for his keen interest in going through the manuscript in spite of his exteremely busy schedule at Yojana Bhawan and writing a Foreword to the volume.

Thanks are also due to S. Kewal Singh, Librarian at Punjab Schol of Economics, Guru Nanak Dev University, Amritsar for preparing the index and to Mrs. Madhu Sharma for expert secretarial assistance. Mr. Arun Kumar Verma, my publisher, also deserves special thanks for his keen interest in its publication and bringing it out at a short notice.

Dr. Gurusharan Singh Kainth

Punjab Agricultural University
Kheti Bhawan, Amritsar 143 001
July 16, 1996

LIST OF TABLES

LIST OF FIGURES

LIST OF ANNEX TABLES

CONTENTS

CHAPTER I

THE PROBLEM OF SUPPLY RESPONSE IN AGRICULTURE

One of the most striking feature of agricultural development in India is the rather rapid evolution of policy, which has responded in the pragmatic manner to the failure of the past and rising needs of the future. It may be argued that by the standards of low income and hungry people, the evolution has been too slow. Under-nutrition and malnutrition are the twin problems of the dietary habits. The people in under-developed areas scramble for food and to stay alive. They are perpetually tired, weak and vulnerable to many diseases. Since their under-nourishment or malnourishment begins from their infancy, it creates permanently depressing and irremediable effects on the society. But compared with many contemporary developing countries, Indian policy reflects more reality than dogma and compared to progress in the early development stages of many new development countries, the pace has been rapid. In India, achievement in the agricultural sector will continue to be a determining factor in the accomplishment of plan targets for many years to come specially in the wake of liberalisation and globalisation of the Indian economy. Further in order to industrialize the economy, we need a substantial improvement in our nourishment, efficiency and raw materials for the industry. This again depends on the rate of improvement in our food and agricultural production. There is, therefore, the need to frame and implement policies which lead to substantial expansion of food and agricultural production.

The tale of Green Revolution in India is primarily about the spectacular increase in foodgrain production, especially of wheat which was initiated with the release of semi dwarf high yielding varieties of Mexican origin. No other major crop in India, except rice, has achieved as high growth rate as wheat production. Punjab, the food basket of India is more closely identified as the 'seat of wheat revolution' than any other state of the country. The high intensive agriculture being practiced by the farmers in Punjab was evident from the fact the geographically the Punjab land area was only 1.5 per cent of the national land area, but its cultivated area was 3 per cent of the total cultivated area and its cropped area 4.3 per cent. Production-wise state was contributing 11 per cent of foodgrain output in the country, 22 per cent of cotton and 10 per cent of milk and milk products. Among cereals crops, rice and wheat are the two major crops extensively grown in the state. Besides the recovery of sugarcane was as high as 10.8 per cent. The cane cultivation in Punjab was sustaining 17 sugar factories.

While all this indicated its immense agriculture production potential, the present agricultural practices may prove suicidal for its population. The warning has been sounded that the entire area might turn into desert if agriculturists continue to over exploit the underground water resources. It was mis-fortune that there were too many tubewells and the water table had started coming down at an alarming rate. Moreover, over exploitation of this potential was adversely affecting the environment. Against this such a high rate of production was accounted for by the institutions working in full harmony: research institutions which made available the new technology, a well responsive robust peasantry and a committed administration ensuring all necessary resource inputs at the right time. Agriculture being the backbone of Punjab economy, the prosperity of farmers is synonymous with the prosperity of the state. The success or otherwise of programmes of agricultural development, in turn, depends largely on the way farmers react to such programmes. In a free market economy price policy could be considered as one of the potent instrument to affect farmer's decisions regarding resource allocation. However, for making the price policy an effective instrument for inducing desired changes in resource allocation under different crop enterprises, it is essential to know how farmers react to different price changes.

The evaluation of a new technology does not necessarily imply its automatic adoption. According to Schultz (1968) three economic requirements, viz., (i) an efficient system of pricing; (ii) a supply of high pay off agricultural inputs; and (iii) the development of sources of these high pay off inputs, must be fulfilled to realize the economic possibilities open to the developing countries.[1] Hallar and Wortman (1969) also laid emphasis on the role of an assured and satisfactory price along with other marketing facilities for the farmer for his crop at the time he plants and before he mortgage his future.[2] Ultimately, out of several measures of public policy, which directly or indirectly influence the farmer's decision, price policy comes out to be a most important instrument having a bearing on the resource allocation decisions of the farmers. It is in this context that the supply response analysis become very important as a guide to formulation of suitable price policies for accelerating production of different crops.

The degree of price responsiveness is basically an empirical question. There are, however, well known and serious difficulties in measuring the degree of responsiveness of produces to price changes. They arise mainly from the difficulties in approximating theoretical formulations of functional relationship to observe real world situations. These difficulties are further compounded because of the time lag between changes in production capacity and changes in output. Despite these problems, the phenomenon of price changes of agricultural commodities and its impact on supplies of the commodities has been a topic of investigation in several studies in the recent past.[3] Due to the time lag between changes in agricultural production capacity and changes in output, in any attempt to measure the price responsiveness, the functional relationship should ideally be worked out between planned output and expected prices.[4] However, the possible discrepancy between planned and realised output and non-availability of any kind of data about planned output except acreage under particular crop have forced the scholars in the field of supply response, to treat acreage (planted or harvested) as a proxy for planned output. On the other hand the necessity of using expected prices has, in turn given rise to many postulates about the ways in which farmers formulate their expectations about the future prices. In the formulation of dynamic models, economists have often to consider expectations about future economic variables and in developing their theories and required to hypothesise how these expectations are formed. Formation of expectations about prices is of considerable relevance in the stability of market equilibrium.[5] By now Nerlove's[6] reformation of Cagan's[7] adoptive price expectation model has become a standard tool for estimating supply functions.

The subject of farmer's responsiveness to economic variables has gained importance during the past few decades in both under-developed and developed countries. Developing economies need to understand the supply phenomena in order to implement policies for raising the output to a level which can provide for adequate human nutritions for the growing population and also to promote general economic developments. Even in developed countries, understanding of the supply phenomena is of crucial importance for controlling surplus, for raising farm incomes and resource productivity.[8]

Objectives

The present study is an attempt to examine the supply behaviour in Punjab Agriculture under changing farm technology. Specifically, the objectives of this study are:

1. To examine the behaviour of rainfall at the selected stations.
2. To examine the nature of growth in the output of selected crops.
3. To measure the influence of certain weather factors on area and yield of selected crops.
4. To examine the supply response to various economic and non-economic factors for selected crops.

Nature of Crop-weather Relationship

Production of a particular crop in any region is the product of area under the crop, and its yield per unit of area. Area and yield can each be expressed as a function of a number of controlled variables (i.e., variables which are directly controlled by farmers and policy makers) like fertilizers, pesticides, irrigation, prices, etc., and an uncontrolled set of variables 'W', comprising various climatic factors like rainfall, temperature, run of dry days, humidity, day length, etc. Therefore, any variation in yield and or area can be conceptually partitioned into two components—one arising out of the controlled variables, and the other from the uncontrolled variables. According to some unknown law of nature, the variables in 'W' vary and thus cause unexpected variations in yield and or area, and hence in production. The various climatic factors enter into the production of a crop in a given region as exogenous variables—they are determined or fixed by nature. The productivity of the various controlled variables in the system differ over space and time because magnitude of exogenous or nature determined variables vary. The year to year variations in agricultural production can broadly be classified into two set of factors viz., "Controllable" and "Uncontrollable". The former category is man made while the latter is made by nature.[9] In a traditional agriculture the latter set of factors play a dominant role while in modern agriculture, the former set is more important. The physical frame of reference in agriculture, is however, not static—it can be influenced through technological advances. Theoretically and conceptually, it is possible to practically eliminate the unexpected variation in production due to natural factors (e.g. on an experimental plot water, temperature, day length, humidity, wind velocity, etc., can be kept at desired levels through artificial techniques). This idealized proposition is, however, difficult to realize when we consider the production function of a region, or for that matter, even a plot on the cultivator's field. Physical factors appear as constraints on the production surface. Their basic differences from region to region lead to differences in land use and cropping pattern.

Soil is the staging area for plant growth, and at any location its quality is itself as expression of the climate. The biosphere over the soil depends on the climate, and the basic soil productivity of an area is a function of the interaction between the biosphere and the climate over years. This basic soil productivity at any location is not subject to any year to year changes, but undergoes a very slow drift over a longer period. For all practical purposes, it is static and cannot be changed. Similarly, when we speak of a region or an area, the supply of sunlight, temperature, water in the form of rain, etc., cannot be altered except by the nature. Their presence or occurrence becomes inevitable event to the farmers. However, people can influence nature through technological advances. The technology of agriculture in any country, at any period, is function of the desire to influence the various natural factors which affect production.

There are, however, frequent references to the fact that technology in agriculture has advanced to such a level that weather no longer has any significant impact on agricultural production. In fact, an opinion is gaining currency in some circles that with the rapid technological changes occurring in agriculture, the studies on crop-weather relationship will not lead to any useful understanding of the phenomenon of year to year change in agricultural production. That this is not correct can be understood from a small example. It is generally argued that in areas with assured irrigation, the influence of weather particularly rainfall on agricultural production is almost negligible. To the extent the supplementary irrigation facilities are available to augment the deficient rainfall, the above argument is valid. But if the rainfall is very heavy in a particular year, a significant part of the crop may be destroyed and ultimately result in lower production. Recent year was one of the severe test for Indian agriculturists with a capricious nature laying bare the strength and weakness of farm economy. Unprecedented drought and floods in the *kharif* season and unseasonal rain at the *rabi* harvest time have left an agricultural economy that is bruised but not battered. Further the drought of 1978-79 dramatically illustrates the extent to which Indian agriculture still depends on the vagaries of nature despite the improved farming methods.

Thus the interactions between direct and indirect influences of weather and between weather and non-weather factors in agriculture are so complex that at whatever level of technology our agriculture may operate, the hold of weather on agricultural production will continue to be a force to be reckoned with.

The need for a study of some measures of the influence of weather on crop acreages and yields is useful from a number of view-points. The first and the foremost is that it helps in improving analysis and estimation of economic relationship in the real world situation. Such studies by measuring the influence of weather would help remove an important obstacle in the way of supply analysis projections. Secondly, it helps in formulating an estimate of a particular crop a few months in advance of the actual harvest of the crop. If the relationship between the play of weather factors, at different stages of crop growth, and yield rates are known the size of the crop can be anticipated in advance which is of immense use to the Government and trader to plan about procurement, distribution, prices, imports etc., of the crop. Another important use of such studies is that they help us to isolate the effects of developmental measures aimed at increasing the agricultural production; and thus in a way assess the progress attributed to planned agricultural development policies. In the present study, therefore, an attempt has also been made to establish the crop-weather relationship. As there are number of *a priori* reasons to believe that acreage and yield of a crop respond differently to the same weather conditions, the present study analysed separately the effect of weather on changes in area and yield of the selected crops.

Supply Behaviour Reviewed

Peasant responsiveness to economic incentives is a crucial element in determining success or failure of programmes designed to improve both domestic standards of living and the balance of payments. This is even more important in case of developing nations which essentially need balanced programmes in terms of resources available as well as targets sought. Thus from early sixties empirical work on supply response started in India and other developing countries. In the past two decades a large body of literature on the application of econometric techniques for the determination of supply response to various economic and non-economic factors, which vary in magnitude from region to region and with the nature of crop, has been developed.

This section reviews the work on the response of agricultural production to changes in economic and non-economic factors in the context of Indian agriculture. No detail critical examination of these models is attempted. The objective is to examine the procedure adopted by these earlier investigators which would help in our attempt to specify and develop an alternative path.

The past studies on supply response are classified into two following categories:

1. Studies relating to crop weather relationships.
2. Studies relating to supply response to economic and non-economic factors.

STUDIES RELATING TO CROP-WEATHER RELATIONS

The empirical work on crop weather relation is largely on the All India Co-ordinated Crop-Weather Scheme (AICWS) data launched by Indian Council of Agricultural Research (ICAR) in 1945. Under this scheme specialized meteorological observatories were set up for "systematic recording of crop-weather observations on paddy, wheat and jowar at a network of selected experimental farms throughout the country." These observatories were later extended to sugarcane and cotton as well. However, earlier to this, the main centre of research on crop-weather relations has been the division of agricultural meteorology of the Indian Meteorological Department (IMD).[10]

Reviewing the status of this work Sarker reported: "Some tentative crop-weather relationships have been established with respect to the crops at different crop-weather stations by applying statistical methods. Response of a few crops in yield to distribution of rainfall during the life cycle of the crop has also been obtained. Similar studies in relation to other meteorological parameters like maximum temperature, sunshine and humidity are in progress."[11]

Some of the early attempts at analysing AICWS data used relatively simple techniques to get an idea of yield-weather relations. For instance, Mallik (1958) examined nine years' data for three crops at the Dharwar Research Station and found that in the two years when the wheat yields were very low due to a rust attack, the number of hours of bright sunshine days during November was abnormally low. From this he concluded that "unseasonal rain and cloudy conditions created conditions favourable for several rust attacks."

On the other hand, in jowar, on the basis of a comparison of rainfall during the growing season in two years of very poor harvests, he argued that "jowar crop at Dharwar is rather susceptible to excessive rain during the growing period." Using a similar approach it was further suggested that the "spell of cloudy and rainy weather extending over 3 consecutive weeks during the growing season of cotton appears to create conditions favourable for pests like shoot borer and red-cotton bug."[12]

In another paper Mallik (1958) attempted a more elaborate analysis of ten years' data relating to Jowar from five stations. He postulated, partly for lack of any other basis, that optimum amount and distribution of rainfall during the growing period of *Kharif* jowar is approximated by the amount of rainfall and its distribution (in each of 12 weeks prior to ear-emergence). He then estimated the correlation coefficients between (a) height and yield; (b) the percentage of the deviation of actual weekly rainfall during the growing period in each year from the rainfall in corresponding weeks of the 'optimum' year (i.e., the year of the maximum yield in the sample) and the percentage deviation from the maximum height; and (c) deviation in rainfall from the year of optimum yield and deviations from optimum yield using a similar procedures as in (b). He found the correlation coefficient of (a) to be significantly positive (0.59), and of (b) and (c) to be significantly negative (–0.42 and –0.51 respectively). The correlation coefficient of yield and rainfall on height was also significantly negative (–0.35). Apart from the fact that there is no explicit a *priori* basis for the relation sought to be estimated, the analysis, as the author recognizes, is limited and tentative as it does not consider the effects of other meteorological variables or the effect of rainfall and other meteorological factors between ear-emergence and harvest, and because it is based on pooling of observations from different stations with varying agro-climatic conditions and possibly different varieties of each crop.[13]

A subsequent paper [14] attempted a rather more elaborate analysis of data for cotton from some 12 stations. Here again, the stations were pooled into two groups (on the basis of rainfall in the reproductive period) to get a sufficient number of observations for studying correlations between (a) different growth features and yield; and (b) meteorological factors and some growth factors. In both groups of districts, the selected growth features (germination rates, number and height of plants at first picking, number of branches, total number of flowers and of bolls at first picking) were found to explain more than 92 per cent of the yield variations. Of the meteorological variables, the correlation analysis revealed that for one group of stations (i) greater rainfall during the four weeks preceding the branching period, as well as during the period, and more rainy days during branching increased the number of branches; (ii) greater rainfall and more rainy days during boll formation reduce the number of bolls available per picking; and (iii) higher maximum temperatures during boll formation tend to increase the number of bolls for picking. In the second group of district the relations are much less clear; Greater sunshine during the four weeks prior to germination seems to increase the germination rate, while more sunshine during boll formation has a marked negative effect on the number of bolls picked.

Sreenivasan (1973) studied the influence of rainfall on the yield of cotton in Madhya Pradesh. The data on rainfall and cotton yield were statistically analysed, pertains to 1925–67 for Khandwa and to 1935–67 for Indore. He used Fisherian technique of regression integral and the technique of selection by significant direct correlation and regression function. Both the techniques are reported to do equally well when judged by the closeness of the predicted to the actual yield levels. The trend in yield found for Khandwa was removed by the Fisherian technique. Five out of six rainfall distribution constants for Khandwa and three out of six for Indore had significant correlation with yield. The regression of rainfall distribution constants before and after the removal of trend were significant for Khandwa but insignificant for Indore. For both the stations additional rains during growth and boll-formation periods was detrimental to the crop. The method of analyses brought out the sensitive weeks in the elongation through boll formation period of 16 weeks.[15]

In another paper, Sreenivasan and Banerjee (1973) presented the results of a linear multiple regression of the yield of *rabi* Jowar in Raichur on mean maximum temperature, mean minimum temperature, total rainfall and number of rainy days. The multiple correlation coefficient is around 0.54 but no further details e.g., the statistical significance of the coefficient, or the basis for the model chosen are available. The paper then goes on to experiment with the "multiple curvilinear regression" technique developed by Ezekiel and Fox (1965) on the same body of data. The method was found to explain much higher proportion of the observed variations in jowar yield in Raichur than linear regression. But unless the underlying model was firmly grounded on theoretical sound and consistent hypotheses derived from the relevant discipline, its utility for understanding crop-weather relations remains doubtful, no matter how high the R^2 value.[16]

Shaha and Banerjee (1975) examined the effect of rainfall, humidity, sunshine, maximum and minimum temperatures, each taken separately (and considering the values of each variable over the entire growing season) on the yield of cotton. They used the "regression integral technique" developed by Fisher (1920) for analysis of crop-weather relations. Some of the main conclusions of their analysis were: while the significance of the effect of rainfall, sunshine and maximum temperature on yield could not be established clearly, there was indication that "more sunshine during the crop period and more rainfall upto the middle of January were beneficial; more relative humidity during elongation and branching period was useful for crop"; and "minimum temperature plays the most important role in explaining about 72 per cent of the total variation in cotton yield.[17]

Sundaresan (1980) conducted a study to analyse the impact of rainfall on crop yield of cumbu in Kovilpatti block of Tirunelveli District. A functional analysis was restored to by fitting multiple regression. The yield of cumbu, the major crop of Kovilpatti block was taken as a function of the annual rainfall, percentage of area irrigated to gross cropped area of cumbu and time trend. The results indicated that 41 per cent of the variation in cumbu yield could be explained by the variables considered. It revealed that an increase of rainfall by one millimeter would, *ceteris paribus* increase the yield of cumbu by 1.12 Kg per hectare. likewise if the percentage of area under irrigation to gross cropped area were to increase by one per cent, the yield would increase by 55.02 Kg per hectare.[18]

Mallik (1980) attempted to examine the influence of some weather factors, namely, rainfall and temperature on area and yield of wheat, rice, maize and groundnut. Keeping in view the wide spatial differences observed in weather variables, study was conducted on small geographical unit. Accordingly, the districts selected for the study were Ludhiana, Patiala, Amritsar (all from Punjab) and Karnal (Haryana). The study utilises the time series data on area, yield, monthly rainfall and average monthly temperature for the period from 1950-51 to 1977-78. The study come to the conclusion that pre-sowing season rainfall was an important determinant of the area sown under wheat in all the four districts under study. Similarly, in the case of rice crop also, the pre-sowing season rainfall significantly affects the area sown under this crop. In case of maize crop, the study of which was confined only to Ludhiana district, the pre-sowing season rainfall significantly influence area allocation to maize. However, the results obtained for groundnut did not speak of any significant influence of the selected weather factors on area sown.

The results of the crop yield-weather relationship reveal that the aridity index (constructed by encompassing rainfall and temperature in a single variable) for the month of January significantly affects the yield of wheat crop. Rainfall during the different growing months of rice crop significantly affect the crop yield. In case of maize yield, it was the temperature during the months of August and September rather than rainfall which affects the crop yield. Groundnut yield was affected by both rainfall and temperature during the growing period. These results are important in view of the fact that almost the entire area under the studied crops was irrigated. The significance of rainfall in affecting the area sown under the crop and its yield under predominantly irrigated conditions imply that the available irrigation is not sufficient enough to overcome the deficiency of rainfall, and that available irrigation supplements rainfall and was not its substitute. Thus the general belief that technological innovations such as irrigation can ensure the freedom from the clutches of the vagaries of weather does not seem to be entirely true.[19]

Rao (1983) using multiple linear regression technique made an estimation of wheat yields for Punjab State and some of its important districts, namely, Amritsar, Patiala, Ludhiana and Ferozepur based on climatic variables, synergistic heat moisture stress and technology. Rao concluded that wheat yields during the period from 1960 to 1980 increased nearly three times due to advances in technology. November maximum temperature, January minimum temperature, March mean temperature and Synergistic heat moisture stress during December were the factors that effect the final crop yield. The State model was used to estimate the district yield and vice versa with a fair degree of accuracy. The models were able to estimate the wheat yields within 10 per cent for 4-year independent data.[20]

In another study Rao with Dudhane (1984) using multiple regression analysis concluded that rainfall, clouds and minimum temperatures during sowing, elongation, tillering and flowering stages are the important weather parameters that affect the final wheat yield over Himachal Pradesh during 1951 to 1967. Lowest wheat yield was associated with poor rainfall, low clouds amount and appreciable fall of minimum temperature over the state. The important weather systems responsible for the differential weather features was the western disturbances.[21]

Agrawal et al. (1984) studied the effects of weather variables on rice yield at different stages of crop growth and forecasted rice yield for Puri district using climatic variables.

The weekly weather variables used were maximum and minimum temperature, relative humidity (at 7 and 14 hrs.), total weekly rainfall, number of rainy days, and sunshine hours. The results of individual effects of weather on rice yield indicated that the crop reacts differently to climatic variables during different stages of its growth. Above average maximum daily temperature had beneficial effects during early weeks of initial growth, and at ripening stage. Detrimental effects were observed during later weeks of the initial and reproductive stage of the crop while during active and lag vegetative phase effects were fluctuating. Pronounced beneficial effect was observed during 10th week. The effect of increase in minimum temperature was beneficial during initial, lag vegetative and reproductive stages, detrimental during ripening stage and fluctuating during active vegetative stage. Above average relative humidity and rainfall had beneficial effects during ripening phase of the crop in general. The effects of rainfall and increase in number of rainy days were fluctuating upto lag vegetative, beneficial during reproductive and detrimental during ripening stage of the crop. Effects of increase in sunshine hours were detrimental during initial growth, active vegetative and reproductive stages, fluctuating during lag vegetative and beneficial during ripening stage of the crop.

Studies on joints effects of maximum temperature and sunshine hours revealed that the effects of maximum temperature on yield was effected positively/negatively by increase in sunshine hours during the weeks when the individual effect of increase in sunshine hours on yield was beneficial/detrimental. The effect of sunshine hours on yield increases in magnitude with increase in maximum temperature throughout the crop season. Forecast model has been developed using weekly weather data of partial crop season. Results indicated that the model using data on maximum temperature and sunshine hours upto 12th week (i.e., third week of August) was appropriate for forecasting rice yield as it explained 91 per cent variation in yield. This suggests that forecasting of rice yield was possible for Puri district 2½ months before harvest. The model based on data upto 1978 has been used for forecasting yield for 1979. The yield forecast was obtained with 11 per cent deviation from actual yield.[22]

STUDIES RELATING TO SUPPLY RESPONSE TO ECONOMIC & NON-ECONOMIC FACTORS

This section concentrates mainly on the empirical studies which deal with supply response of agricultural production to various economic and non-economic factors in the chronological order in the context of Indian agriculture. Since most of the study has used Nerlovian model, we begin with the summary of this model.

Marc Nerlove (1958) had attempted to estimate the elasticity of supply for corn, cotton, and wheat in the U.S.A. His work had dealt primarily with the role the farmers' expectations of future prices play in shaping their decisions as to how many acres they should devote to each crop.[23]

Nerlove tried to identify the price variable appropriate in a supply function with an observable variable or several observable variables. He had devised a model relating expected "Normal" price to past observed prices. If P^*_t be the expected normal price during period t, P_t be the observed price during period t, and B be a constant called the coefficient of expectation. The model had been presented as:

$$P^*_t - P^*_{t-1} = B(P_{t-1} - P^*_{t-1}); \quad 0 < B \leq 1 \quad \ldots (1)$$

If the observed output X_t is a linear function of P^*_t and a random residual term u_t, mathematically, we may write:

$$X_t = a_0 + a_1P^*_t + u_t \quad \ldots (2)$$

Nerlove had shown that the observed output X_t may be written as a function of the past observed prices as in equation (3) and as a function of the lagged observed price and lagged observed output as in equation (4):

$$X_t = a_0 + a_1[BP_{t-1} + B(1 - B)P_{t-2} + \ldots] + u_t \quad \ldots (3)$$

$$X_t = a_0B + a_1BP_{t-1} + (1 - B)X_{t-1} + u_t - (1 - B)u_{t-1} \ldots (4)$$

Nerlove had also considered the difference between the long and short run equilibrium output and the difference between the long and short run elasticity of supply. He had developed a model relating the short run equilibrium output X_t to the long run equilibrium output X^*_t.

Let E be a constant called the coefficient of adjustment. The model developed may be written:

$$X_t - X_{t-1} = E\,(X^*_t - X_{t-1});\ 0 < E < 1 \qquad \ldots \quad (5)$$

Let it be assumed that X^*_t is a linear func on of P_{t-1} then equation (2) may be written as:

$$X^*_t = a_0 + a_1 P_{t-1} + u_t \qquad \ldots \quad (6)$$

Nerlove had also dealt with the problem of adjustment to a changed level of prices over time. The estimates of the elasticities of supply of corn, cotton, and wheat presented by Nerlove were substantially higher than what had heretofore been obtained. Estimates of the elasticity of cotton acreage to expected normal price range between 0.7 and 4.5; estimates of the elasticity of wheat acreage with respect to expected normal price range between 0.9 and 1.2; estimates of the elasticity of corn acreage with respect to expected normal price range between 0.2 and 0.4.

After the work of Nerlove, most of the empirical studies have made use of the Nerlovian adjustment model notwithstanding the fact that these studies differ in respect of the shifter variables included in the regression equation and the nature of the price variable used.

Raj Krishna (1963) estimated the supply response of 11 major crops on the basis of time series data of undivided Punjab using Nerlovian adjustment lag model. His analysis revealed that of the major crops (cotton american, cotton desi, maize, sugarcane, rice, bajra, jowar, wheat, gram and barley) all except jowar had positive short run elasticities varying from as low as 0.08 and 0.09 in case of wheat irrigated and bajra to the medium magnitudes 0.23 to 0.34 and 0.31 in case of maize, sugarcane and rice upto 0.59 and 0.72 in the case of cotton desi and american respectively. The corresponding long run elasticities of the crop in above order wheat to cotton ranged from 0.15 to 1.62. The study also showed that besides prices, other non-price variables such as yield, rainfall and irrigation were also important, at least for few crops. The coefficients of adjustment estimated from the equations also indicated the rapidity of adjustment of the acreage of crop by farmers which could be favourably compared with that of United States.[24]

Jakhade and Majumdar (1964) examined farmer's response to changes in price parities between jute and rice in West Bengal, Bihar and Assam during the period of 1950–1964. They made use of Tabular analysis in working out price parity ratios from their average pre-sowing prices, and their relationships with the acreage under the two crops. A plausible relationship was obtained between the parity price movement and the shifts in acreage suggesting thereby that parity price ratios could be used, within certain limits, as instruments for regulating shifts in acreage from one crop to the other crop in the desired direction.[25]

To the growing volume of literature on producers' response to price stimulus, Dharam Narain's study (1965) is a well-authenticated and thought-provoking contribution. He had attempted a rigorous exposition of almost all important aspects of the hypothesis concerning producers' behaviour in response to price changes in the context of peasant agriculture. The complex and intricate theoretical issues involved in assessing the role of basic economic principal of profit motive in the decision-making process of the Indian farmers had been thoroughly examined. The second part of his study was devoted to a graphical comparison of year-to-year variations in acreage, under six crops, namely, cotton, jute, groundnut, sugarcane, rice and wheat, with those in their prices and proceeds per acre. The prices of competing crops and data on rainfall had also been juxtaposed. While the graphical method of analysing the complex relationship between prices and acreage has the merit of simplicity, it lacks precision. It has also not been possible to quantify the significance of the influence of different variables, which as econometric approach would have attempted.

Narain had concluded that the Indian farmer was significantly responsive to prices. He had, however, pointed out a striking difference of behaviour between cash and food crops. Similarity between the behaviour of areas under cash crop and the prices of those crops, corrected with reference to changes either in an index of agricultural prices or in the prices of specific competing crops, in the majority of the cases was so striking that prices emerges as a decisive consideration to the farmer with the areas he sows for those crops. The study reveals that the Indian cotton grower was significantly responsive to commercial stimuli as his counterpart engaged in highly commercialized agriculture in the U.S.A

In the case of foodgrain crops, however, rainfall assumes the status which prices does in the case of cash crops. The study also demonstrates the role of prices in foodgrain areas. The basis for expecting the play of prices in foodgrain areas was, however, firmer when foodgrain compete with cash crops rather than with rival foodgrains in a given area. On the basis of his selective findings relating to rice area in the jute region of Bengal and a cotton-growing tract in Madras, to wheat area in a part of the sugarcane-growing region of United Provinces, to jowar area in relation to area under groundnut in Madras, and to area under millets, generally in the cotton-growing regions of the country, Narain had proved that, in specific regions at least, prices exerts a significant influence on the variations of foodgrain areas.[26]

Gupta and Majid (1965) estimated the responsiveness of sugarcane and paddy acreage in Deoria district of Eastern U.P., during the period 1949-50 to 1961-62. They considered farm harvest prices of paddy and guaranteed minimum price of sugarcane lagged one year as the prices relevant in their analysis. By removing the trend through method of link relatives, they had computed the link relatives of relative prices and the relative acreage on an year-to-year basis. The analysis suggested that a continuous expansion of acreage under sugarcane in relation to paddy from year-to-year may not altogether be a function of variations in the relative prices of sugarcane and paddy from year to year.

The importance of this study lies in that they had collected data for sugarcane and paddy for the year 1961–62 to compare net monetary gains from the cultivation of an acre each of sugarcane and paddy. The net monetary return per acre from sugarcane was 4 times high as that for paddy and this provided an explanation for increase in sugarcane acreage relatively to that of paddy. The development of cooperative marketing system for sugarcane and the various facilities and inducements provided to the producer for sugarcane cultivation, e.g., the fixation of a guaranteed price and its announcement in advance by the government and the assurance of a market at that price, also helped to accentuate the expansion of sugarcane acreage. With the help of data collected through the survey method the authors had analysed the response of the selected producers to variations in prices and to marketing policies.[27]

John (1965) studied the behaviour of price, area and output under rice and sugarcane in U.P., during the period 1954-55 to 1962-63 by working out annual rates of change in prices, area and output, both in absolute as well as in relative terms. He observed that while the relative price of sugarcane to that of rice had been declining, its area and output relative to that of rice had been increasing. He concluded that the existence of a wide gap between per hectare returns of sugarcane and that of rice was responsible for this erratic behaviour.[28]

Kahlon et al., (1965) examined the influence of wheat price and other factors on acreage allocation of wheat in Punjab. They used time series data from 1950–51 to 1961–62 to run regression analysis. The results revealed that wheat acreage responded to change in the prices of wheat in a significant way over the period of study. Cross elasticity of wheat acreage relative to gram price was significant at 5 per cent level. Rain-fall and time factor indicated positive influence on crop acreage but the coefficient were non-significant.[29] In another study Kahlon alongwith Grewal (1966) studied the factors responsible for variation in cane supplied during the period 1960-61 to 1964-65 in the Punjab. They found that area under sugarcane in 1963-64 declined considerably as compared to 1960-61 and 1961-62 because of faulty purchase policy followed by sugar factories. Secondly, 'gur' prices were relatively higher. It was indicated that significant increase in cane acreage during 1964-65 was due to unusual competition from gur industry for cane supplies.[30]

Krishna and Rao provide us with other response evidence in two papers. In these two studies the subject of investigation was limited to wheat in Uttar Pradesh during the period 1950-51 to 1962-63. In their first paper (1965) twelve price formulations based on wholesale price had been used to explain the change in wheat acreage during the reference period. The result of the study indicated that the influence relating to acreage response changes drastically, depending on the estimate of the expected price used. They concluded that the "pre-conceived motions of both the schools of thought i.e., those hypothesizing high degree of supply response and those hypothesizing negligible supply response, can be sustained if the results for different models are looked at individually."[31]

The second paper (1967) of the author was an extension and refinement of their first paper. Nine alternative price expectation models and six different response equations had been used in the study. The results of the study indicate that acreage under wheat had been fairly elastic to changes in relative price of wheat and substitute crops during the period of study. The analysis suggested that the three year average presowing prices of wheat deflated by three year average of pre-sowing prices of substitute crops along with the relative yield of wheat and rainfall were the most important factors influencing farmer's decision on area allocation among wheat and substitute crops. They concluded that traditional regression models for estimating the supply response coefficient, if properly specified, can give as satisfactory, if not superior, results as those obtained by using adjustment lag models of the Nerlovian type. Most of the short run elasticities derived from the Nerlovian models had been found to be non-significant. The long run elasticities obtained from some of the traditional model specifications had been found to range between 0.60 and 0.72. These estimates compare favourably with those obtained from the adjustment lag models of Nerlovian type. The Nerlovian type model had been found to have edge over the traditional ones in so far as the proportion of variation in wheat acreage is explained by the explanatory variable.[32]

Behrman (1968) modified Nerlovian model by including risk aversion in the estimation of supply response. Farmer's rational conduct, following a risk aversion hypotheses would imply that subjective price and yield probability distribution, farmers would seek to maximise expected return for a given level of variance in the expected returns, which is a measure of risk. Additional variables like lagged yield, rainfall were also considered in the structural equation for the period 1937 to 1963 for major crops in Thailand. In addition to a significant response to price in the total supply of rice, the author found even larger response to price in the marketed supply of the crop. For major four Thailand crops, rice, cassava, corn and kenaf, short run price elasticities were greater than estimated for crops in other countries.[33]

Kaul and Sidhu (1971) attempted to relate acreage under different crops in Punjab to changes in relative profitability of the crop and lagged acreage and introduced the concept of variability in the relative profitability models. The authors maintain that the inclusion of variability in the form of coefficient of variation rather than standard deviation in the model, improved the explanatory power of the equations for paddy, groundnut and desi cotton, but the elasticities did not change materially except in the case of paddy.[34]

Maji, Jha and Venkataraman (1971) estimated the supply and demand elasticities for wheat, maize and rice in Punjab for the period 1948–49 to 1965–66. They incorporated a risk variable in the Nerlovian lag adjustment model with a view to obtain some ideas regarding farmer's response to the variation in prices. While the absolute price proved to had significant impact in case of wheat,. the risk factor was not significant but after its addition the magnitude as well as the significance of price coefficient increased. In case of maize relative prices rather absolute prices had significant impact and higher value of adjustment coefficient suggested the absence of rigidity which inhibits speedy adjustment in the long run. In case of rice, only relative price coefficient was significant and that too on the traditional model only. They also suggested that qualitatively better estimates emanate from the dynamic models than the traditional models.[35]

In another paper Maji and Jha (1971) tested the application of Cobb-Webb theories to explain fluctuations in sugarcane in four districts of Bihar for the period 1934-35 to 1964-65. Two types of models, namely, traditional Cobb-Webb models and dynamic supply model had been used to indicate cyclical functions in area with respect to prices. Both the models showed that Cobb-Webb was of the convergent type, the traditional model suggesting a two year periodicity of the price cycle and the dynamic version indicating cycles of 4–5 years' duration. The coefficient of adjustment found in acreage response function was 0.7306. Relative price and risk variables were not included in the study.[36]

Parikh (1971) used the data from Dharam Narain's book "*The impact of price movements on areas under selected crops in India: 1900–1939*" to test the hypothesis that farmer's do not respond to changes in profitability until subsistence needs are satisfied by employing various models for rice, wheat and other crops. He observed that the farmer's respond more favourably to non-price factors such as rainfall, varietial improvements, expansion of irrigation and total area under crops in the previous year than to changes in prices. Regarding farmer's response to price changes, no definite conclusion emerged but the statistical evidence presented to support the subsistence theory.[37]

Tyagi (1971) attempts at explaining the possibilities of building up models of farmers expectancy behaviour by analysing the nature and basis of farmer's future price expectations and for using these models for generating the expected prices which could be used explicitly for finding the functional relationship between the expected prices of a crop and the acreage allotted to that crop. The data pertaining to expectations, including socio-economic variables, had been collected by conducting a survey based on a comprehensive and well-structured schedule. The sample of the study was the Garhmukteshwar block of the Meerut district in Western Uttar Pradesh. Tyagi had analysed the foundation of expectations and the expectancy behaviour, i.e., how different farmers formulate their expectations about future prices in general. He had tried to identify the variables which enter into the formulation of future price expectations and the way in which the effect of different variables is aggregated to arrive at the expected price. The study concluded that farmers expect prices which they were likely to get for their produce products in the form of ranges of prices within which the future prices were expected to fall. The expectations about the future prices were mostly based on the observed behaviour of price movements during the preceding two years though other (non-price) variables such as governmental action, political changes, crop prospects, etc., also influence expectations to some extent. Farmers generally expect prices to move in the direction of the expected trend by an amount which was a function of the past observed changes. However, the rate at which prices were expected to increase was greater than the rate at which they were expected to come down.

For measuring the degree of price responsiveness, a functional relationship between the expected future price and the acreage under that crop had been established. To compare the results of the two approaches of estimating supply response, one using expected future prices and the other using actual realised prices of some previous period, the elasticity of supply response had been estimated from different functions using both types of prices. The results of this study provided positive support to the *a priori* hypothesis, viz, "Farmers in under-developed countries respond quickly, normally, and efficiently to relative price changes."[38]

Bal and Rao (1973) estimated short and long run projections of area under wheat in the Punjab state for the period 1950-51 to 1968-69. They used standard deviation of prices in the preceding three years as the explanatory variable instead of prices and to overcome the problem of non-normal distribution of the dependent variable. They divided the entire equation with the standard deviation of the prices. They claimed better results with the new models. Assuming prices to be constant they made the short run projections of acreage under wheat using time trend as the only independent variable in the alternative exponential and sigmoid type of functions.[39]

Singh and Rao (1974) carried supply response study based upon the data on the Indian Virginia Tobacco for the period 1940-41 to 1967-68. They found that the inclusion of dummy variables in the model led to an improvement over the usual estimation procedure of aggregate relationship, as it enabled to quantify the impact of inter-regional characteristics on the magnitude of variations in the level of total acreage in each of the major producing areas of Virginia Tobacco. Besides price, the impact of price risk variables was found to be negative and also reasonably significant. With another variables, the disease and pest control incorporated in the estimating model, the results showed a negative though non-significant impact of this variable.[40]

Ray (1981) attempts to analyse the effects of weather and prices on the output of major food and commercial crops of India. The results of the analysis had amply justified the reliability and usefulness of constructing crop-wise weighted rainfall indices as proxy variables for weather. Analysis made with the crop-wise area and output data from 1951-52 to 1974-75 had convincingly shown that significant part of the fluctuations around their respective trend lines could be explained by the computed rainfall indices. The effect of rainfall on yield was found to be more pronounced than on area. Also, its impact on the commercial crops was generally of lower order than on food crops. An interesting aspect of the analysis was the computation of weather-adjustment growth rates in area, production and yield for different crops in India. It was observed that, even after accounting for weather and slower rate of expansion in area, yield of all the crops except wheat had shown a tendency to decline in the post-1965 period.

The question was then posed: Is weather the only variable responsible for fluctuation in area and output and, what are the underlying forces responsible for generating trend-like movement in crop-wise area and output? Specifically, the interest was to examine, the extent to which the fluctuations in crop output could be explained by the price changes observed during the period. The problem involved in the examination of the above issue was two-fold: first, to construct for each crop, a national index representing the price received by the producers and second to build an adjustment model appropriate in the Indian context. Crop-wise annual indices of prices received and prices paid by the producers were constructed from the month-wise index number of whole-sale prices, with weight derived from certain assumptions made on the marketing behaviour of the producers. The model developed and used in the analysis postulated a two-tier adjustment mechanism for the observed area under the crop at the national level. The model was, however, over-identified and the structural parameters of the model could not be estimated by the ordinary least square technique. Even then, tentative results obtained through OLS procedure provided meaningful interpretation of crop-wise area and output variations observed during the period covered.

Evidently, many more experimentations would be needed before one could fully develop an appropriate model for forecasting the output of different crops in India with reasonable degree of confidence.[41]

Lahiri and Roy (1985) tried to investigate the supply response of rice in India after incorporating rainfall in a careful manner. There seems to have been a specification error in the earlier studies with rainfall specified to have a monotonic impact on supply. The detrimental impact of excess rainfall and floods was ignored in earlier research. The new specification of the gamma type had the added advantage that the 'optimum' rainfall requirement can be estimated from the structure. It is also possible to construct a crop-specific rainfall index to capture the 'pure' effect of rainfall after accounting for the impact of changes in other variables such as prices and irrigation. This index allows the precipitaion of different years to be compared and the index can be used to project supply.

Incorporation of this non-monotonic 'gamma' type impact of rainfall had led to the rejection of adaptive expectations and partial adjustment in the Nerlovian tradition in the case of Indian rice. Expectations were found to be static and farmers appear to be adjusting fully to changes in relative prices within the year. The estimated price-elasticities of Indian rice turn out to be fairly low with respect to acreage changes but significantly high price-elasticities emerge for yield and output. Switching out of rice into an alternative crop was not easy perhaps because the agronomic characteristics of rice-growing regions tend to be specific to rice. Rice farmers were, however, clearly responsive to changes in their relative input-output prices which were, consequently, important instruments for policies geared to influencing rice yields and levels of output.[42]

Kainth (1986) made an attempt to examine the acreage response of wheat to economic and non-economic factors in Indian Punjab under changing farm technology. But, within the state there exist acute variations in agricultural development across different districts. To capture the magnitude of this varied development, different districts of the state were grouped into two sub-groups on the basis of wheat productivity for the year 1979-80 through 1981-82. The district of Ludhiana was selected to represent relatively developed districts and Amritsar to represent under-developed districts. The principal source of secondary data was the *Statistical Abstract of Punjab*—an official publication of the Economic and Statistical Organisation of the Punjab Government. The time reference of the study was 1961-62 through 1982-83. However, district-wise data were not available for some of the parameters from 1961–62. Therefore, the time reference of the disaggregated (district-wise) analysis was 1965-66 through 1982-83.

To test the hypothesis relating to the factors influencing the farmer's acreage allocation, the Nerlovian adjustment lagged model was used. The result of the analysis revealed that in the process of making the area decisions for wheat cultivation, the magnitude of the area held in the preceding year had proved to be the most significant factor. Its main cause, as is evident from the general psychology of the farmers in the state, was that they follow tradition in making decisions about raising different crops. They did not easily decide to reduce the area under the crop in comparison with the area they sowed in the preceding year. In this connection, their own requirements of food, ignorance about determining the appropriate cropping pattern owing to illiteracy and other traditional and social standards adopted by the farmers and the lack of suitable competing crops, were the possible causes. The value of the Nerlovian coefficient of adjustment indicates a very slow adjustment on the part of the farmers. It takes almost twelve to twenty years for 95 per cent of the effect of the price change to be realised in the study area. Thus various techno-institutional and subjective factors seem to influence a great deal the decision making of the farmers in Punjab regarding wheat acreage. The positive acreage irrigation response in the state of Punjab points to the fact that farmers will shift to wheat crop if irrigation is improved.

The result obtained in this study lead to important implications that seem to be relevant from the point of view of policy formulation. First, the varietial improvements in the wheat crop and the adoption of the varieties by the farmers having comparative advantages (technological, institutional and others) will be important for maintaining production at the desired levels. Second, the price and risk factors will need to be taken care of by appropriate measures in order to provide the necessary incentives to the producers; these policies must indeed go hand in hand with policies related to production technology. Third, the result of the study indicate a positive response of land resources allocation to prices. This means that the farmers can find it possible to make adjustments on the acreage allocation under wheat cultivation through the manipulation of relative prices of wheat and substitute crops. In order to bring about an effective adjustment in acreage allocation, the support prices for various crops must be announced well before the sowing season and the prices thus announced, should carry a long run guarantee. This policy will not only enable the farmers to plan their production programmes better but might also help to correct the inter-commodity imbalances to some extent.[43]

In another study Kainth (1987) reported that per hectare yield had played and is expected to play a greater role in increasing production of crops. Hence there will be less and less leverage left to acreage in case of relative change in prices, particularly in a state like Punjab, where most of the culturable wastes have already been brought under plough. The results of the study indicated that quantity of fertilizer used per hectare had positive influence on yield of wheat. Similarly, the influence of percentage area under high-yielding varieties of wheat and percentage wheat area irrigated was positive and highly significant. On the other hand quantity of fertiliser used was being significantly influenced by wheat fertiliser-price ratio and also by the percentage area under high-yielding varieties and percentage wheat area irrigated. The influence of all three variables was positive and highly significant. The explanatory power corrected for degree of freedom of all the fitted model showed high degree of goodness of fit. Wheat-fertiliser price significantly influenced the quantity of fertiliser used. This inturn substantially affected the wheat yield. Percentage area under HYVs and percentage wheat area irrigated also significantly influenced the yield of wheat during the period under reference. Therefore, wheat yield could be stimulated through the spread of high-yielding varieties, extension of irrigation facilities and through positive price policy. The effect of bad weather could be mitigated through the spread of irrigation infrastructure and timely integrated pest management.[44]

Majority of the studies referred to above have approximated planned output to which the supply function actually refers, by acreage planted or harvested. They presume that allocation of non-land resources will be influenced by changes in relative prices in the same manner and to the same extent as allocation of land. Validity of this presumption is not yet empirically established. Moreover, it is the responsiveness of total output to price which is important to policy makers. Total output is determined by the yield and area planted. Thus, elasticity of planned output will be the sum of elasticity of planted area and of planned yield (output per unit of area), assuming that interaction between them will be negligible and hence can be ignored. The higher the elasticity of yield, higher will be the under-estimation of elasticity of planned output by the elasticity of planted acreage.

Again, most of the studies have either omitted the rainfall effect on supply or have introduce it in a monotonic fashion—either simple linear or logarithmically linear. There seems to have been a specification error with this. It does not recognise the detrimental impact of excess rainfall, floods as well as drought.

The present study is a modest attempt to incorporate these issues in later chapters. We estimate supply functions for both acreage and yield; output response is derived as a reduced form by linking area and yield equations. Rainfall is introduced into the supply-response functions in a non-linear fashion that recognises the detrimental impact of floods as well as droughts.

REFERENCES

1. T.W. Schultz, *Economic Growth and Agriculture*. McGraw Hill, New York, 1968.
2. J.G. Hallar and S. Wortman, "Expanding Food Production in Hungry Nation: The Promise and Problems" in Clifford M. Hardin (Ed.) *Overcoming World Hunger*, Prentice Hall, New Jersy, 1969.
3. A brief resume of these studies are presented in the section of supply response reviewed.
4. The farmers are required to commit a sizeable amount of inputs at the beginning of crop season to various crops, the output of which will come after the end of the season, i.e., he is required to plan on the basis of the prices which he anticipates to get for the crop products at the end of the crop season.
5. O.H. Brownke and W. Grainer, "Farmer's Price Anticipations and Planning," *Journal of Farm Economics*, Vol. 31, 1949.
6. Cagan Philp, "The Monetary Dynamics of Hyper-inflation" in *Studies in the Quantity Theory of Money*, (Ed.) M. Friedman, Chicago University Press, Chicago, 1954.
7. Mahesh Chand, "Agricultural Terms of Trade and Economic Growth," *Indian Journal of Agricultural Economics*, Vol. 13, No. 1, January–March 1958.
8. Earl O. Heady, "Uses and Concepts in Supply Analysis," in Earl O. Heady et al. (Eds.), *Agricultural Supply Functions—Estimating Techniques and Interpretation*, Iowa State University Press, Ames, Iowa, 1961.
9. The level of almost all the controlled inputs employed by the farmers in the production process are also significantly influenced by weather. We will consider such variations in production due to weather-induced changes in the levels of controlled inputs like land, irrigation, fertilizer, etc., as made by nature.
10. L.A. Ramdas, *Crops and Weather in India*, Indian Council of Agricultural Research, Monograph, New Delhi, 1960.
11. R.P. Sarker, *Meteorology at the Service of Agriculture*, (Mimeo.) 1977.
12. A.K. Mallik, "Effect of Weather on Crop Growth and Yield at Government Experimental Farm at Dharwar," *Journal of Biological Sciences*, Vol. 1, No. 1, 1958.
13. A.K. Mallik, "Height and Yield of Kharif Jowar in Relation to Rainfall During Vegetative Growth," *Indian Journal of Meteorology and Geophysics*, Vol. 9, No. 4, 1958.
14. A.K. Mallik, P. Jagannathan, G. Rama Rao and J.R. Banerjee, "Preliminary Studies of Crop-Weather Relations," *Indian Journal of Meteorology and Geophysics*, Vol. 11, No. 4, 1960.
15. P.S. Sreenivasan, "Influence of Rainfall on the Yield of Cotton in Madhya Pradesh," *Indian Journal of Agricultural Sciences*, Vol. 43, No. 9, September 1973.

16. P.S. Sreenivasan and J.R. Banerjee, 'Studies on the Forecasting of Yield by Curvilinear Technique", *Indian Journal of Meteorology and Geophysics*, Vol. 24, No. 1, January 1973.
17. S.K. Shaha and J.R. Banerjee, "Influence of Rainfall; Humidity; Sunshine; Maximum and Minimum Temperature on the yield of Cotton at Coimbatore", *Indian Journal of Meteorology and Geophysics*, Vol. 26, No. 4, 1975.
18. Sundaresan, *A Study on Variation in Rainfall and its Impact in Drought Prone Area of Tamil Nadu*, Centre for Agriculture and Rural Development Studies, Tamil Nadu Agricultural University, Coimbatore, Research Report/Ag. Economics/2/80, 1980.
19. R.P.S. Malik, *Influence of Rainfall and Temperature on Crop Production*, Research Study No. 80/4, Agricultural Economics Research Centre, University of Delhi, Delhi, 1980.
20. G. Appa Rao, "Estimation of Wheat Yields over Punjab Using District and State model", *Mausam*, Vol. 34, No. 3, 1983.
21. G. Appa Rao and S.N. Dudhane, "Weather feature Associated with Abnormal Wheat Yield in Himachal Pradesh", *Mausam*, Vol. 35, No. 3, 1984.
22. Ranjana Agarwal, R.C. Jain and M.P. Jha, *Effect of Weather on Rice Yield and its Forecast*, (Puri District), Division of Crop Forecasting Methodology, Indian Agricultural Statistics Research Institute, New Delhi, 1984.
23. Marc Nerlove, *The Dynamics of Supply : Estimation of Farmer's Response to Prices*, The John Hopkins Press, Baltimore, 1958.
24. Raj Krishna, "Farm Supply Response in India-Pakistan: A Case Study of Punjab Region", *Economic Journal*, Vol. 73, No. 2, September 1963.
25. V.M. Jakhade and M.A. Majumdar, "Response of Agricultural Producers to Price—A Case of Jute and Rice in India," *Indian Journal of Agricultural Economics*, Vol. 12, Nos. 3 & 4, 1964.
26. Dharam Narain, *Impact of Price Movements on Area under Selected Crops in India: 1900–39*, Cambridge University Press, Cambridge, London, 1965.
27. S.C. Gupta and A. Majid, *Producers' Response to Changes in Prices and Marketing Policies*, Asia Publishing House, Bombay, 1965.
28. P.V. John, "Responsiveness of Relative Area, Output of Sugarcane and Rice to Change in their relative Prices in U.P.," *Indian Journal of Agricultural Economics*, Vol. 20, No. 1, January-March 1965.
29. A.S. Kahlon, S.S. Johl and H.N. Dwiwedi, "Structure of Farm Prices in the Punjab," *Indian Journal of Agricultural Economics*, Vol. 20, No. 1, January–March 1965.
30. S.S. Grewal and A.S. Kahlon, "A Study of Fluctuations in Cane Supply Industries in Punjab," *Indian Sugar*, Vol. 16, No. 5, 1966.
31. M.S. Rao and Jai Krishna, "Price Expectation of Acreage Response for Wheat in Uttar Pradesh," *Indian Journal of Agricultural Economics*, Vol. XX, No. 1, January–March 1965.
32. Jai Krishna and M.S. Rao, "Dynamics of Acreage Allocation for Wheat in Uttar Pradesh: A Study of Supply Response," *Indian Journal of Agricultural Economics*, Vol. XXII, No. 1, January–March 1967.

33. J.R. Behrman, *Supply Response in Under-developed Agriculture: A Case Study of Four Major Annual Crops in Thailand*: 1937–63, North Holland Publishing Co., Amsterdam, 1968.
34. J.L. Kaul and D.S. Sidhu, "Acreage Response to Price for Major Crops in Punjab—An Econometric Study", *Indian Journal of Agricultural* Economics, Vol. 26, No. 4, October–December 1971.
35. C.C. Maji, D. Jha and L.S. Venkataraman, "Dynamic Supply and Demand Models for Better Estimation to Projections—An Econometric Study for Major Food-grains in the Punjab Region," *Indian Journal of Agricultural Economics*, Vol. 26, No. 1, January–March 1971.
36. C.C. Maji and D. Jha, "Cobb-Webb Phenomenon to fluctuations in Sugarcane Acreage in North Bihar," *Indian Journal of Agricultural Economics*, Vol. 26, No. 4, October–December, 1971.
37. A. Parikh, "Farm Supply Response: A Distributed Lag Analysis", *Bull. Oxf. Ins. Econ. Stat.*, Vol. 33, No. 1, Oxford, 1971.
38. D.S. Tyagi, *Farmer's Expectations of Future Prices and Their Impact on Acreage Allocation: A Case Study in Supply Response in Western Uttar Pradesh*, Ph.D. Thesis, Sardar Patel University, Vallabh, Vidyanagar, 1971.
39. H.K. Bal and D. Raghava Rao, "Long run and Short run Projections of Area under Wheat in the Punjab State," *Journal of Indian Society of Agricultural Statistics*, Vol. 25, No. 2, 1973.
40. R.D. Singh and D.R. Rao, "Determinants of Supply Behaviour: The case of Virginia Tobacco in India", *Artha Vijayana*, Vol. 16, No. 3, 1974.
41. S.K. Ray, "Weather, Price and Fluctuations in Agricultural Production," *Indian Economic Review*, Vol. 16, No. 4, 1981.
42. Ashok Lahiri and Prannoy Roy, "Rainfall and Supply Response: A Study of Rice in India," *Journal of Development Economics*, Vol. 18, No. 1985
43. Gursharan S. Kainth, "Acreage Responses of Wheat in Punjab," *Journal of Social and Economic Review*, Vol. 3, No. 3, July–September 1986.
44. Gursharan S. Kainth and A.S. Kang, "Price Responsiveness of Indian Punjab Wheat Yield—A Macro View," *Agricultural Situation in India*, Vol. 45, No. 12, March 1987.

CHAPTER II

THE RESEARCH MATERIAL

In order to meet the objective outlined in the preceding chapter, both macro (state)-and micro (district)-level statistical data were used. Statistical and econometric analysis are not possible in the absence of quantitative data. Therefore, an important step in research work is the collection of statistics relating to the various aspects of the problem with reasonable accuracy. The basic data used in this study are: (a) area, output and yield statistics, (b) prices statistics, (c) irrigation statistics, and (d) meteorological statistics.

The Area and Crops

The study is confined to Punjab state—one of the most agriculturally advanced state of Indian Union.[1] But within the state there exist acute variations in agricultural development across different districts. Moreover, as is well known, the intensity of different climatic variables differ considerably even in proximate areas. Further, even, within an area, different crops responds differently to the same climatic conditions. As such no useful purpose will be served, if the supply behaviour are examined at an aggregate level of a country or a state. Likewise, it will be worthless to measure the crop-weather relationship by relating the weather factors to aggregate agricultural production. In addition, since the effect of different climatic variables differ at different biological crop's phases, it will not be fair to relate the average climatic conditions, obtained during the entire period of crop growth, with the crop production.

In order to capture the varying magnitude of development and climatic conditions, the present study considered district as the geographical units of study. The underlying assumption is that the climatic variables at a particular point of time do not vary much over the district. Further, it is assumed that the data collected at the weather recording observatory located at a particular place in the district represents the weather conditions of the entire district. Though the assumption does not strictly hold true, however, in the absence of the availability of data on a more detailed geographical classification, the assumption, it is believed, is not irrelevant. The consideration of district as the geographical unit of study has also been advocated by the National Commission on Agriculture.[2]

The different districts and the various crops taken up for the present study are reported in Table 2.1. The selection of these districts has mainly been governed by the uniformity in availability of meteorological data and other aspects of crop production viz., area and productivity (yield per hectare). These crops taken together accounted for nearly 80 per cent of the gross cropped area.[3]

Area, Output and Yield Statistics

Since area sown for raising a crop is a better judge of the producer's decision than area matured or harvested, we have used statistics of sown area. Reliable information on area under

Table 2.1: Selected districts and different crops taken up for the analysis

Crop(s)	Relatively Developed Districts	Relatively Less Developed Districts
Wheat	Ludhiana	Gurdaspur
Rice	Ludhiana	Amritsar
Cotton Desi	Bathinda	Ferozepur
Cotton American	Bathinda	Ferozepur
Sugarcane	Patiala	Gurdaspur
Rapeseed and Mustard	Amritsar	x

each crop is available from cadastral surveys conducted each year by the *patwari*—the village accountant.[4] Further, time series estimates of planned production are conspicuous by their absence. Therefore, we have used time series estimates of realized output which are available and dependable.

The traditional method of estimating yield was the *anawari* system, which involves the concepts of normal yield and a condition factor. This method is highly subjective and the yield estimates depends upon the subjective judgement of the *patwaris* regarding normality of rainfall and other yield affecting conditions.

Significantly enough, the crop-cutting survey (CCS) replaced the traditional eye-estimation method in Punjab during 1960-61 for wheat and paddy. Sugarcane has been covered since 1965-66, rapeseed and mustard since 1973-74 and cotton since 1974-75 under CCS. Estimates of production are arrived at by multiplying the area under the crop by the average yield per unit of area, latter being obtained, generally speaking, through crop-cutting experiments.

The crop cutting experiments are designed by a properly drawn sample. The sampling design is a multi-stage stratified sample. Crop-cutting experiments consists of selection of a few villages in each tehsil of a district by random sampling method, selection of a few fields in each selected village randomly and consequently selection of a plot of 1/80th an area in each selected field at random. Average yield per hectare is estimated from the crop harvested on such plots.

The main source of area, production and average yield of the selected crops in the selected districts are the Director of Land Records and Director of Agriculture (Punjab). These statistics, however, are regularly published in *Statistical Abstract of Punjab*—an annual publication of Economic and Statistical Organization of Government of Punjab. These statistics are recorded from the different issues of these abstracts.

Irrigation Statistics

Water is as necessary for the growth of the crop as blood is for human body. Two types of irrigation statistics are needed for our analysis—net area irrigated and the gross area irrigated. These statistics are regularly collected by the Director of Land Records and reproduced in *Statistical Abstract of Punjab*. As such, irrigation statistics were collected from these different issues.

Price Statistics

The information of prices is required for generating data on responsible prices, i.e., prices used by the farmers in their production decision. Two types of price statistics are required viz., (i) output prices, and (ii) input prices.

Output prices that should be used in such an analysis is the weighted average of several prices received by different farmers throughout the years; weights being the quantities of the crops disposed off at different points of time in a particular year. But a long time series of such prices are not available. Two types of prices are generally available—farm harvest prices (FHP) and wholesale prices (WP). Farm harvest prices are the average of weekly wholesale prices in certain selected markets of the district during the periods closely following the harvesting of various crops. The post-harvest period includes six to eight weeks[5] after the normal harvesting period. The wholesale prices are based mainly on reports received from market committees of the important market centres and pertains to the calendar year. Farm harvest prices are considered more suitable for the present purpose because these prices are better indicators of the prices received by the farmers than wholesale prices.

The main limitation of the farm harvest prices stems from the fact that farmers sell a portion of the surplus later in the lean period and difference in prices received from that portion is not taken care of. But the farmers sell major chunk of their produce in immediate post harvest period.[6] In addition, the storage of farm produce does involve costs and these costs normally may be commensurate with the increase in prices during the lean period. Moreover, different government agencies are procuring grains at the same price throughout the year. It can, therefore, safely be assumed that the farm harvest prices represents the true prices. Data on FHP has been collected from different issues of the *Statistical Abstract of Punjab*.

Input Prices

Farmers use a large varieties of chemical fertilizer to supply the three primary plant nutrients—nitrogen (N), phosphorus (K_2O) and potassium (P_2O_5). Input (fertilizer) prices that might have been used in the analysis was the weighted composite index of fertilizer price, with the quantities of the various fertilizer actually consumed as weights. But the historical quantity data were not available. In addition, out of three primary plant nutrients, majority of farmers applied only nitrogenous fertilizer, which is one of the most important plant nutrients. As such we make use of composite index of nitrogenous fertilizer only. For this purpose, two nitrogenous fertilizers—calcium ammonium nitrate (CAN), and the urea—were selected as these were the most commonly used fertilizers. A simple average of these two nitrogenous fertilizers was taken to represent the composite fertilizer price in terms of nutrient availability. Following conversion rates were used to convert the market price in terms of nutrient price:

Fertilizer	Conversion Factor (%)
Urea	46
Calcium Ammonium Nitrate (CAN)	25

The above conversion rates are based upon the availability of nitrogen nutrient in the respective fertilizer. Data regarding fertilizer prices has been collected from *Fertilizer Statistics*, a publication of the Fertilizer Association of India (FAI), New Delhi.

Meteorological Statistics

Obtaining suitable data for an expert measure of crop-weather relation is extremely difficult. An ideal measure would require the data from the experiments conducted under controlled conditions in glass houses where as many variables as desired are controlled. By this approach we can get information on the response of plants to one or more meteorological factors. It is, however, extremely difficult even by this approach to stimulate the variations and interacting effects of all climatic factors. Another though relatively less desirable data source would be the data obtained from the standardized field experiments where excepting for weather, all other variables are held constant. But such a time series data under Indian conditions are not available. Even comparable meteorological data are available only on a few select variables. Taking into consideration, the above data limitations we make use of the available historical data on meteorological variable as collected by Indian Meteorological Department (IMD). The details of these meteorological variables selected are given in the following paragraphs:

Rainfall data have been complied from the daily statements issued by the Regional Meteorological Centre, New Delhi. Normal rainfall figures are based on data of State rainguage stations, except for Patiala where data of observatory stations only are available. These normals are calculated from the available data collected at the respective stations upto the end of 1940, except for Patiala where the period is 1945 to 1950.

Temperature figures are also based on the daily statements issued by Regional Meteorological Centre, New Delhi, and are recorded at 08.30 hours, Indian Standard Time (IST). Highest and lowest temperature are the highest and the lowest temperature on a particular day in a month. Mean temperature is one-half of the average maximum plus average minimum temperature recorded during the month.

Relative humidity is the measure of the amount of water vapours present in the atmosphere relative to the corresponding amount for saturation and are recorded at 08.30 hours, Indian Standard Time (IST).

Reference Period

The time reference of the analysis is twenty three years spanning the period 1971 through 1993. Since the state was reorganised during 1965–66, the study relates to the reorganised Punjab. Further during August 1972, some parts of Ferozepur district had been merged with the then Faridkot tehsil of Bhatinda district to form a new district of Faridkot. Therefore, district-wise data for some of the parameters were available from 1971 only. The choice of the period was, therefore, mainly dictated by the availability of complete and reliable data.

Selection of Months

The effect of different climatic variables differ at different biological crop's phases. It will, therefore, not be fair to relate the average climatic conditions, obtained during the entire period of crop growth with the crop production. We have, thus, related the crop area and yield to the relevant meteorological variables affecting them at different biological phases. Depending upon prevalent cultural practices in the study region,[7] the selected months for each of the crop studied are as reported in Table 2.2

Limitations

The main limitations of the present study have already been discussed. These limitations have crept into the study by the very nature of the available data. Needless to say, the validity of our conclusions are contingent upon the accuracy of these data. It is, however, hoped that these will not seriously impair the results of the study.

Table 2.2: Selected months for crop-weather relationships for different crops taken for the study

Crop(s)	Selected months for	
	Area Response	Yield Response
Wheat	September–November	October–April
Paddy	May–July	June–October
Cotton	March–May	April–November
Sugarcane	January–March	February–December
Rapeseed & Mustard	August–September	September–December

REFERENCES

1. G.S. Kainth and R.S. Bawa, *Economic Development and Structural Changes: An Empirical Investigation*, Inter-India Publication, New Delhi, 1985.
2. G.O.I., *Report of the National Commission on Agriculture Climate and Agriculture*, Vol. 4, Ministry of Agriculture and Irrigation, New Delhi, 1976
3. E.S.O., *Statistical Abstracts of Punjab—1993*, Economic and Statistical Organisation, Government of Punjab, Chandigarh. 1994.
4. I.C.A.R., *Hand Book on Methods of Collection of Agricultural Statistics in India*, Indian Council of Agricultural Research, New Delhi, 1959.
5. G.O.I., *Agricultural Statistics in India—Development in Methodologies*, Directorate of Economic and Statistics, Ministry of Food and Agriculture, New Delhi, 1968.
6. G.S. Kainth, *Foodgrain Marketing System in India, Its Structure and Performance*, Associated Publishing House, New Delhi, 1982.
7. The views of the area-agronomists were also taken regarding the selection of different periods for the crops under study.

CHAPTER III

RAINFALL BEHAVIOUR IN PUNJAB

Recent technological break-through in agriculture has helped to make crop husbandry profitable and is expected to estimate monetisation of the rural economy. However, such a frame work lends itself all the more sensitively to elements of uncertainty. Variability of weather factors like rainfall, temperature, humidity, wind velocity, sunlight and so on during different biological phases of the crop production have always constituted a major source of uncertainty in farm decision making. The problem of crop production enterprise organisation and resource allocation has become more accurate with the introduction of new varieties as the farmers have little experience of the response behaviour of these varieties to weather factors and their interaction with farm inputs. Managerial decisions in farming, therefore, cannot afford to neglect the variation in the weather factors. Although assured irrigation is a powerful means to reduce the vagaries of drought and ill-distributed rains, it does not correct all types of adverse effects of weather. There is apparently the lack of studies which had tried to estimate the effect of weather vagaries under-irrigated conditions. Important questions which need answers are: Does agriculture production carry risks of weather under fully irrigated conditions? Can the effect of weather under such condition be estimated? All these questions are greatly influenced by the climatological parameters, more specifically the rainfall. A proper understanding of the pattern of variation in rainfall would help in adjusting crop practices to the vagaries of nature, so that crop losses can be minimized. Recognizing the important role of rainfall in increasing or depressing the crop production, it was, therefore, considered essential to analyse, in depth, the distribution pattern of rainfall, so that effective strategies could be identified.

Punjab have a continental climate with extremes of temperature ranging from 45 to 50 degree celsius in summer to near freezing point in winter. Depending upon the weather conditions, whole of year may be broadly divided into four seasons:

Season	*Months*
1. Winter rains	January–February
2. Pre-monsoon/ Hot weather	March–May
3. South-west monsoon/ Rainy season	June–September
4. North-east monsoon	October–December

The main rainy season starts in July and continues till September. The monsoon season rainfall not only affects the production of *Kharif* crops, but also affects the area and yield of *Rabi* crops by increasing the availability of waters due to retained moisture in the soil. Winter season rainfall, though occasional and small in quantity, also improves the yield of *Rabi* crops. The annual rainfall varies from 400 to 800 mm in plains and from 800 to 1500 mm in submontane districts. Accordingly, Punjab falls in the category of dry areas. Due to large scale exploitation of ground water and surface water resources, the irrigated area works out to be 90 per cent of the gross cropped area. But for the large scale utilization of irrigation potential, Punjab would have become a chronic drought prone area.[1]

A study of rainfall is of interest to meteorologists, agricultural scientists and economists alike. However, uncertainty of rainfall and its impact have to be an essential part of our understanding of natural resources of a region. No regional plan can rest on a firm basis in the absence of a proper knowledge of the behaviour of rainfall. National Commission on

Agriculture has devoted a full volume out of its 15 volume report for the analysis of climate and agriculture. The work relates to rainfall sub-division of the country as a whole. But, we are concerned with district wise rainfall, i.e., with smaller regions. Moreover, we have raised further issues for analysis which could help in a better understanding of rainfall at the state level. Some of the important issues discussed in this chapter are:

1. What is the nature and magnitude of variations in district-wise monthly; seasonal and annual rainfall over a long time series. Is there any relation between normal rainfall and magnitude of variation therein ?
2. What is the probability of rainfall occurrence in different ranges relative to the normal level ?
3. What is the relation between the level of annual rainfall and its seasonal timeliness? What are the salient features of variation in monthly rainfall ?

Probability Estimates for Rainfall

The prediction of rainfall at a particular place and time can be made by studying the behaviour of rainfall of that place over several years during the past. This behaviour is best studied by fitting a suitable distribution to the time series data on the rainfall. The most important statistics of a distribution are the mean and variance. For comparing the variability, they are often combined into the form of a coefficient of variation. The rainfall is predicted with the help of the probability estimates. For studying the variability of rainfall as well as obtaining the probability estimates of rainfall, one has to decide about the suitable period for which the data is to be combined. It could be a fortnight, or a month or the entire season. We have considered the three cases, namely, (a) the annual, (b) the monthly, and (c) the seasonal.

For determining a suitable mathematical function for rainfall distribution, the statistics β_1 and β_2 were calculated and reported in Annex Table 3.17 for annual and seasonal rainfall and in Annex Tablc 3.18 for monthly rainfalls. These estimated values differed significantly from 0 and 3 respectively. Thus, no suitable distribution could be found even after transforming the data to logarithmatic or square root scales. Accordingly, we have restricted our analysis to the nature and magnitude of variations.

Annual Rainfall and its Variation

Quite a few studies have tried to identify productivity or cyclical pattern and see if it is possible to make long or short range forecasting. A consensus of such studies as reported by the National Commission on Agriculture is that, "Rainfall series of stations in the country generally exhibit random nature and have shown no significant trend on the whole. Some of the long rainfall series have been tested for periodicity also but none of significance has been noted."

The average annual rainfall at different selected rain gauge stations (RGS) of Punjab were examined (Table 3.1) for the period 1971 through 1993 and plotted in Figure 3.1. Average annual rainfall fluctuated within a range of 31.83 cm (Bathinda) to 98.99 cm (Gurdaspur). However, the coefficient of variation (CV) varied from 22.83 per cent (Gurdaspur) to 38.84 per cent (Amritsar). A close look at Fig. 3.1 would reveal the following points:

1. a. Rainfall series starts and ends with below normal year in the malwa region of Punjab (Fig. 3.1a)

Table 3.1: Annual rainfalls; variation therein; number of turning points and Z-Statistic at selected stations in Punjab: Average of 1971 through 1993

Station(s)	Estimated Value of			Number of Turning Points (NTP)	Z-Statistic
	Mean (in cms)	C.V.	P.F.		
Amritsar	63.21	38.84	40	15	0.515
Bathinda	31.83	36.92	31	16	1.031
Ferozepur	41.63	33.86	26	14	N.C.
Gurdaspur	98.99	22.83	22	16	1.031
Ludhiana	67.78	31.36	20	13	0.515
Patiala	72.76	35.28	22	15	0.515

P.F. = Probability of Failure, C.V. = Coefficient of Variation.

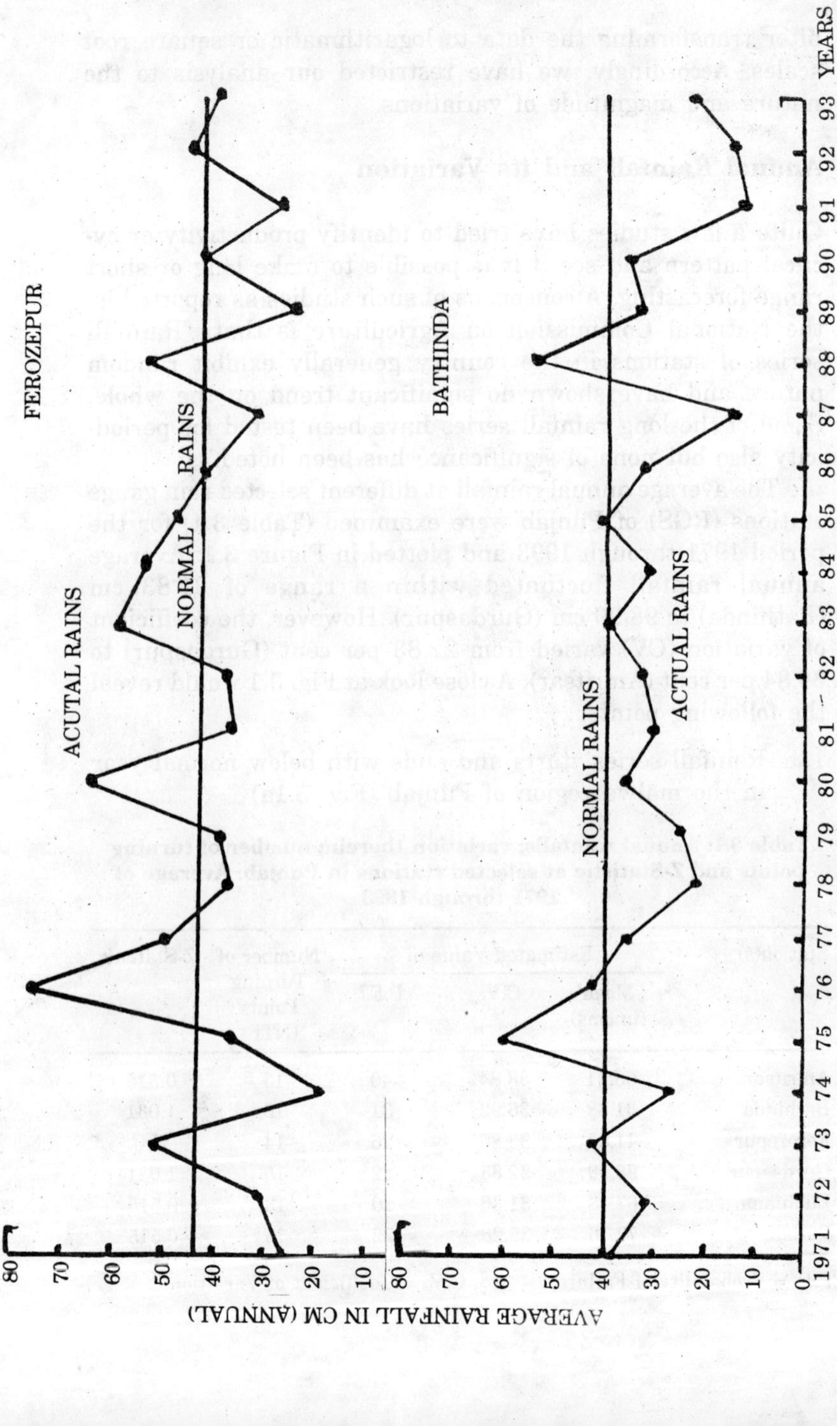

Fig. 3.1a: Average annual and normal rainfall (cm) at selected rain gauge stations of Ferozepur and Bathinda: 1971 through 1993

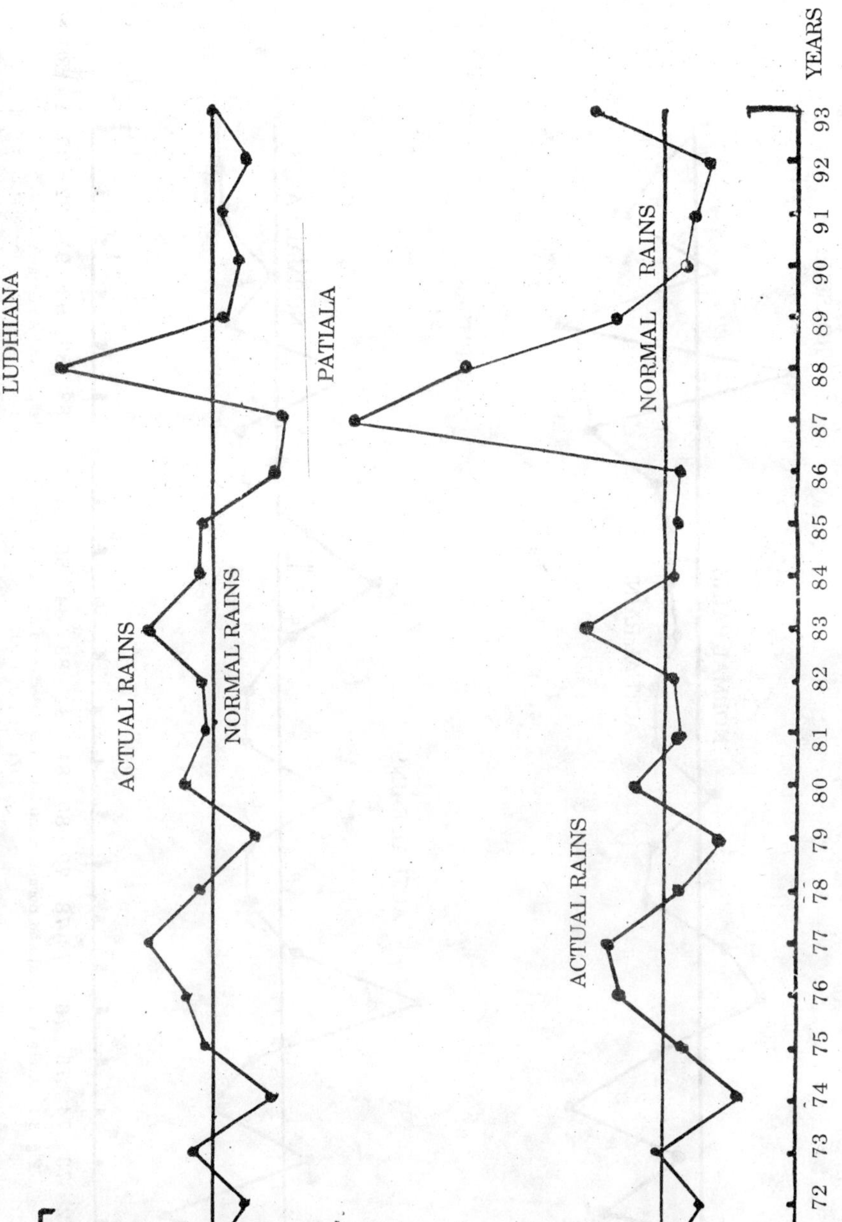

Fig. 3.1b: Average annual and normal rainfall (cm) at selected rain gauge stations of Ludhiana and Patiala: 1971 through 1993

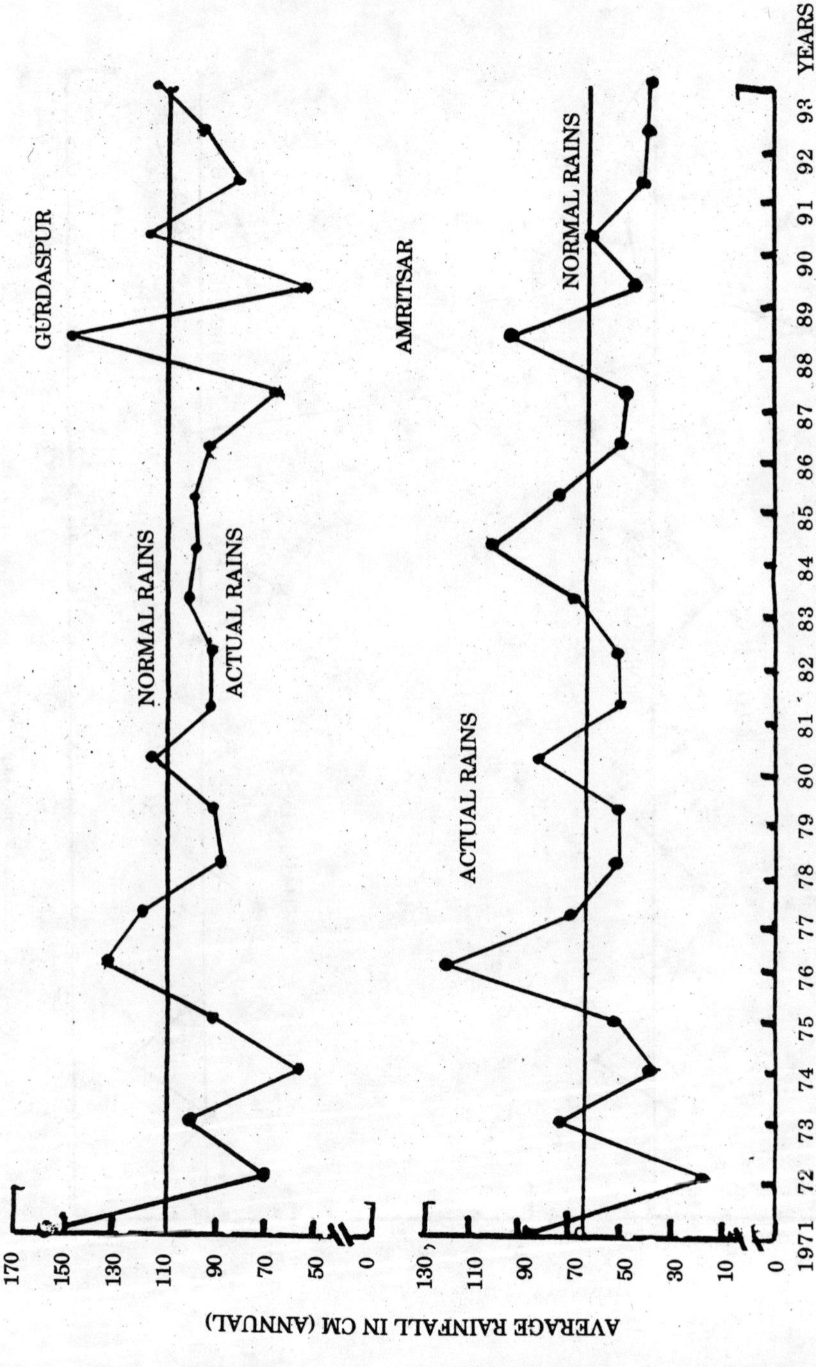

Fig. 3.1c: Average annual and normal rainfall (cm) at selected rain gauge stations of Gurdaspur and Amritsar: 1971 through 1993

b. Rainfall series starts with almost at normal year but ends above the normal year (Fig 3.1b) in the central Punjab.

c. In Majha region of Punjab, rainfall series starts with a good normal year but ends with the below normal in Amritsar and just at normal in Gurdaspur (Fig. 3.1c) that is, Shivalik foot hills.

2. There is no definite trend in the rainfall over the study period—for all the selected rain gauge stations.
3. Although there are periodic fluctuations, no definite cyclical pattern of rainfall is observed. In any case, in a short series of this type, it is difficult to establish the existence of a regular cycle.

The trend in the annual rainfalls at the selected stations in Punjab during the reference period was also statistically examined by using the Z-statistic (Kendall 1958) defined as:[3]

$$Z = \frac{n - E(n)}{S.E.(n)}$$

Where n is the observed number of turning points in the data; and

$$E(n) = 2(n - 2)/3$$

and,

$$S.E.(n) = \sqrt{(16N - 29)/90}$$

N being the total number of observations. The estimates of number of turning points and its associated Z-statistic are presented in Table 3.1. The test gave no evidence of presence of a trend in the annual rainfalls at the selected rain gauge stations of Punjab. The absence of definite trend in the annual rainfalls at the selected RGS of Punjab was confirmed by testing the series against linear time trend. The estimates of linear time trend in annual rainfalls are reported in Annex Table 3.1.

It may, however, be added that efforts at predicting rainfall are continuing. Bean saw some evidence of 'retro-symmetry' in rainfall data of a few areas in India, and argued that if such focal points are traced it could help prediction.[4] According to press reports Dr. M.S. Swaminathan, the distinguished farm scientist has observed that Auto Regressive Integrated Moving Averages (ARIMA) model has proved quite helpful in predicting rainfall.[5]

Nature of Variation

A perusal of the annual rainfall pattern showed that the variation in annual rainfall has been random and exhibit no trend or cycles of regular frequency or amplitudes. In general, the gaps between troughs or severe failure of rainfall in any particular year cannot be predicted from the trends or cycles in the past. Hence it is possible to study only the magnitude of instability in rainfall and the risks associated with it.

Instability can be said to have two dimensions—one is the variability as measured by the coefficient of variation, and second is the nature of probability of distribution itself, that is, whether it is flat, peaked, skewed or simply irregular defying any set pattern. A dependence only on the coefficient of variation would hardly provide a full understanding of the nature of variations in rainfall. Hence in addition to studying the coefficient of variation, the probability (PF) distribution of rainfall was also studied. Annual rainfall is considered to have failed if it is less than 25 per cent from the normal. The number of years of rainfall failure are expressed as percentage of total number of years and reported in Table 3.1. These two measures go together as evidenced from the positive correlation between the two (viz $r = +0.688$). There is, however, no exact one to one correspondence between the two. In general, districts having higher variations in rainfall are having higher probability of rainfall failure.

It is important to know if uncertainty in rainfall as measured by CV and or PF is associated with the level of normal rainfall. Are high rainfall areas also subject to almost the same uncertainty as low rainfall districts. One need not take any relation for granted in this respect. In Punjab, however, inter district picture showed a negative and statistically significant correlation (r = –0.736) between the average annual rainfall and coefficient of variation and also non-significant negative correlation (r = –0.546) with probability of failure and average annual rainfall. This makes it necessary to identify district in various categories in terms of the magnitude of average rainfall and uncertainty therein.

We can see from Table 3.1 that Amritsar Majha region of Punjab have high rainfall and high uncertainty therein. The diametrically opposite group consists Malwa region of Bathinda and Ferozepur[6] districts with low rainfall and relatively low uncertainty. On the other hand, Ludhiana and Patiala of central Punjab have moderate to high rainfall but low uncertainty. Each type of area may call for a different crop pattern and different agricultural strategy.

It is not enough to study failure in terms of an aggressive measure. One has to study the probability or frequency of rainfall in different ranges to see if extreme failure is more significant or moderate failure. Table 3.2 presents the probability distribution of annual rainfall in seven ranges of 20 percentage points of average for the reference period. It can be seen from Table 3.2 that the frequency of extreme deviations beyond 30 per cent of the normal on either side is not high with the only exception of Bathinda where the estimated probability of failure is about 31 per cent. Even under a lib-

Table 3.2: Probability of occurrence (as per cent of total years) of rainfall in different ranges from normal (average) at selected station in Punjab: Average of 1971 through 1993

Station(s)	Ranges of deciles from Normal (Per Cent)						
	Failure/ below Normal			Normal	Excess/above Normal		
	Below -50	-30 to -50	-10 to -30	-10 to +10	+10 to +30	+ 30 to +50	Above +50
Amritsar	4	10	35	13	13	9	9
Bathinda	18	13	26	26	9	4	4
Ferozepur	4	13	26	26	13	9	9
Gurdaspur	x	17	35	26	13	9	x
Ludhiana	x	17	13	35	22	9	4
Patiala	x	9	17	39	13	13	9

eral criterion of failure, the total probability of failure below 10 per cent of the normal has always been less than the total probability of rainfall being in the normal range or above. The maximum probability of failure, thus, defined is in Bathinda. Interestingly, Bathinda is a district which is defined as Drought Prone because of its low average rainfall. Majority of the cultivated area (about four fifths) were canal irrigated.

Gurdaspur, Ludhiana, Patiala and Ferozepur have the distinction of having more than one thirds of the total years within the normal range of rainfall viz., from (–) 10 per cent to (+) 10 per cent. Amritsar and Bathinda has only one fourths of the total years within the normal range of rainfall. Moreover, the major region of Punjab, consisting of Gurdaspur and Amritsar too have higher degree (more than two fifths) of probability of failure. Both these districts claimed equal proportion of net area irrigated by canals as well as by tubewells.

Seasonality of rainfalls

It hardly needs to be emphasised that more than the annual total rainfall, its distribution across different periods of plant growth or in different seasons is important for any particular crop(s). The same amount of rainfall may be obtained in an optimum distribution over various months or in a number of less favourable combinations. Rainfall before sowing, after sowing, after flowering time, at grain formation time etc., affects output in different ways. It is, therefore, essential to take into account the seasonal distribution of rainfall. A study of variation on monthly and or seasonal rainfall and the extent of seasonal untimeliness in each districts can be helpful in agricultural planning of the districts concerned.[7]

Table 3.3 presents the seasonal[8] distribution of rainfall and variation therein. During the reference period, rainy season average rainfall varied from 76.62 cm (Gurdaspur) to 27.39 cm (Bathinda). The coefficient of variation varied between 46 per cent (Amritsar) to 26 per cent (Ludhiana). Apparently, rainfall during south-west monsoon was quite stable and did not vary much across different stations. The diametrically opposite was the picture during north-east monsoon. Average rainfall varied from 1.72 cm (Bathinda) to 6.06 cm (Gurdaspur)

Table 3.3: Seasonal rainfalls and variations therein at selected stations in Punjab: Average of 1971 through 1993

Station(s)	Average rainfall (in cms) during				Coefficient of Variation				Probability of failure in percentage			
	WSR	PMR	SWR	NER	WSR	PMR	SWR	NER	WSR	PMR	SWR	NER
Amritsar	5.31	6.68	49.77	3.59	0.47	0.78	0.46	1.06	25	40	30	40
Bathinda	2.50	2.90	27.39	1.72	0.84	0.34	0.43	0.96	50	55	25	50
Ferozepur	2.65	4.45	32.42	1.83	0.59	0.79	0.38	0.97	40	55	20	40
Gurdaspur	9.16	9.62	76.62	6.06	0.50	0.76	0.28	0.98	30	35	20	70
Ludhiana	5.08	5.33	49.82	3.16	0.61	0.72	0.26	0.93	30	40	20	45
Patiala	5.39	5.42	50.57	3.76	0.63	0.96	0.32	1.18	45	35	20	50

WSR : implies Winter Season Rains
PMR : implies Pre Monsoon or Hot Weather Rains
SWR : implies South-West or Rainy Season Rains
NER : implies North-East Rains

whereas the coefficient of variation varied from 118 per cent (Patiala) to 93 per cent (Ludhiana). Apparently, rainfall is more dependable during rainy season of heavy rainfalls. Probability of failure also showed the similar pattern and varied from 15 per cent (Ferozepur to 30 per cent (Bathinda).

Monthly distribution of rainfalls[9] at the selected rain gauge station are presented in Table 3.4 and plotted in Figure 3.2. Table 3.4 also presents the variations in monthly rainfalls in terms of coefficient of variations and probability of failure of rainfall by 25 per cent or more from the normal of respective month(s). During the month of July, the average rainfall between different RGS of Punjab varied from 28.62 cm (Gurdaspur) to 10.03 cm (Bathinda) and the probability of failure varied from 30 per cent in (Bathinda) to 15 per cent in Ferozepur. During the same period, the coefficient of variation varied between 28 per cent (Gurdaspur) to 66 per cent in Ferozepur. This implies that the rainfall during the month of July was quite stable and did not vary much from place to place. The average amount of rainfall during the month of July was maximum with minimum variation therein, in terms of both the measures viz., coefficient of variation and probability of failure, as compared to other months. This indicates that July corresponds to heaviest rainfall month with good deal of stability. For the month of August, which too happens to be a heavy rainfall month, the average amount of rainfall between the selected RGS varied from 29.50 cm (Gurdaspur) to 10.38 cm (Bathinda). the coefficient of variation, however, varied from 74 per cent (Bathinda) to 35 per cent (Patiala). Likewise the probability of failure varied from 15 per cent (Patiala) to 40 per cent each in Bathinda and Ferozepur. In the month of September, the average amount of rainfall varied from 10.67 cm (Gurdaspur) to 3.54 cm (Bathinda). The coefficient of variation varied from 96 per cent (Amritsar) to 59 per cent (Ludhiana) and the probability of failure from 35 per cent in Ferozepur, Gurdaspur, Ludhiana each to 50 per cent (Bathinda).

On the other hand, it was observed that the months of low rainfall were associated with high value of coefficient of variation and probability of failure therein. For example, during the month of November, the average monthly rainfall varied

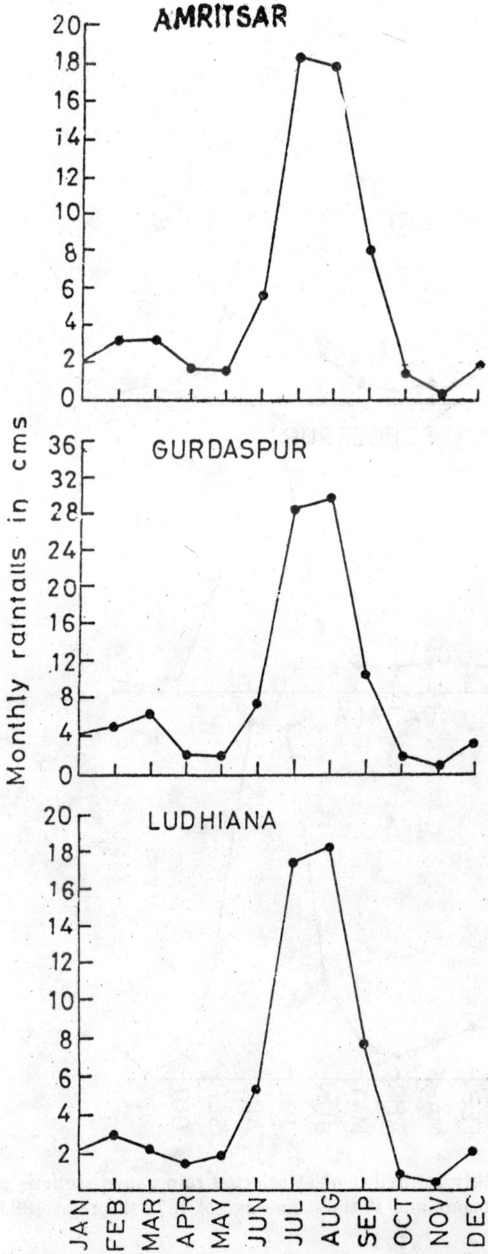

Fig. 3.2a: Monthly rainfall (cm) at selected rain gauge stations of Amritsar, Gurdaspur and Ludhiana: Average of 1971 throught 1993

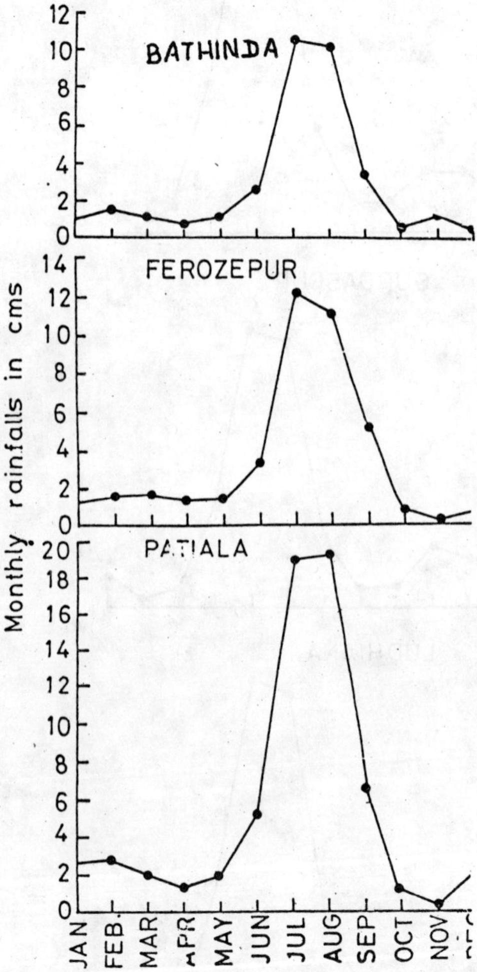

Fig. 3.2b: Monthly rainfall (cm) at selected rain gauge stations of Bathinda, Ferozpur and Patiala: Average of 1971 throught 1993

Table 3.4: Monthly rainfall and variations therein at selected stations in Punjab: Average of 1971 through 1993

Month(s)	Amritsar	Bathinda	Ferozpur	Gurdaspur	Ludhiana	Patiala
			Average Rainfall (in cms)			
January	2.18	0.91	1.12	4.45	2.11	2.56
February	3.13	1.59	1.53	4.71	2.97	2.83
March	3.17	1.08	1.70	5.06	2.07	2.05
April	1.78	0.76	1.35	2.65	1.47	1.38
May	1.73	1.06	1.40	1.91	1.79	1.99
June	5.60	2.85	3.84	7.86	5.75	5.49
July	18.35	10.63	12.14	28.61	17.63	19.02
August	17.90	10.37	11.01	29.50	18.47	19.27
September	7.92	3.54	5.43	10.65	7.97	6.79
October	1.44	0.67	0.83	1.94	0.83	1.16
November	0.24	0.29	0.25	0.64	0.36	0.34
December	1.91	0.76	0.75	3.48	1.97	2.26
			Coefficient of Variation			
January	0.74	1.09	0.98	0.69	0.99	1.04
February	0.62	1.03	0.81	0.54	0.80	0.75
March	1.06	1.51	1.16	0.93	0.95	0.88
April	1.69	1.98	1.66	1.50	1.59	2.25
May	0.97	1.21	1.12	0.74	0.98	1.08
June	0.76	0.73	0.84	0.76	0.58	0.52
July	0.50	0.61	0.66	0.28	0.35	0.46
August	0.66	0.74	0.72	0.39	0.43	0.35
September	0.96	0.75	0.79	0.62	0.59	0.67
October	1.63	1.64	1.49	1.53	0.91	1.16
November	1.56	2.10	1.44	1.92	1.72	2.12
December	1.27	1.84	1.37	1.28	1.22	1.42
			Probability of Failure (in Percentage)			
January	45	55	50	40	50	45
February	20	45	35	30	50	40
March	45	60	35	40	25	45
April	55	60	80	50	50	60
May	65	60	60	30	55	65
June	45	40	35	55	35	30
July	20	30	15	20	25	25
August	30	40	40	25	25	15
September	45	50	35	35	35	40
October	60	50	70	65	45	60
November	60	75	65	55	70	60
December	45	60	50	40	55	60

from 0.25 cm (Ferozepur) to 1.29 cm (Bathinda). The coefficient of variation varied from 212 per cent in Patiala to 144 per cent in Ferozepur. Probability of failure varied between 55 per cent in Gurdaspur to 75 per cent in Bathinda.

Instability, therefore, in terms of the coefficient of variation (CV) and the probability of failure (PF) of rainfall by 25 per cent or more from the normal of respective month(s), is very high for the individual months, particularly for the months which receive scanty rainfalls. There is no strict one-to-one negative correspondence, but in general high rainfall months are also months of relatively more dependable rainfall.

Annex Table 3.4 presents the probability distribution of rainfall in seven ranges of 20 percentage points of average 1971 through 1993 for seasonal and Annex Table 3.5 through 3.10 of monthly rainfalls respectively. During south-west monsoon, we find that the frequency of extreme deviations beyond 30 per cent either way is not high for all the selected rain gauge stations. Diametrically opposite is the picture for the other three seasons, namely, winter rains, pre-monsoon and north-east monsoon. Gurdaspur has the distinction of having more than half of the total years within the normal range of rainfall (from –10 to +10 per cent) during the rainy season. Gurdaspur and Patiala have the distinction of heavy 45 per cent of the total years within the normal range (from –10 to +10 per cent) of rainfall during the month of August. Gurdaspur also has 40 per cent of the total years within the normal range of rainfall (from –10 to +10 per cent) during the month of July.

Variability in Relative Shares

During the study period about 78 per cent of the average annual rainfall was received during south-west monsoon (rainy season) and 5 per cent was received during north-east monsoon. The winter season received nearly 8 per cent of the total rainfalls. The rainfall during the winter season did not benefit the crop production much in view of the strong irrigation infrastructural base in the state. The rainfall during hot weather (pre-monsoon) was 9 per cent of the annual totals.

Table 3.5 presents the average share (1971 through 1993) of monthly rains in the annual total rainfall and its coefficients of variation. The pre-monsoon showers start in April itself, but it is really from June onwards upto September that rainfall is significant at all the rain gauge start. In all the rain gauge stations (RGS), either July or August claims the honour of highest share in the annual total. On the other hand, rainfall was lowest at all RGS in the month of November. It is also interesting to note that rainfall during the month of December to March was also significant.

Table 3.5: Monthly shares (per cent) of rainfalls in annual rainfalls and its coefficient of variations at selected stations in Punjab: Average of 1971 through 1993

Month(s)	Amritsar	Bathinda	Ferozepur	Gurdaspur	Ludhiana	Patiala
			Percentage (average) share			
January	3.34	2.64	2.71	4.39	3.33	3.93
February	4.79	4.61	3.70	4.64	4.69	4.34
March	4.85	3.13	4.11	4.99	3.27	3.15
April	2.72	2.20	3.26	2.61	2.32	2.12
May	2.65	3.07	3.39	1.88	2.82	3.05
June	8.57	8.26	9.29	7.75	9.07	8.42
July	28.08	30.80	29.36	28.20	27.81	29.21
August	27.39	30.05	26.63	29.07	29.14	29.59
September	12.12	10.26	13.13	10.50	12.57	10.42
October	2.20	1.94	2.01	1.91	1.31	1.78
November	0.37	0.84	0.60	0.63	0.57	0.52
December	2.92	2.20	1.81	3.43	3.10	3.47
			Coefficient variation			
January	77.97	107.33	111.73	66.68	88.78	91.30
February	59.84	111.18	61.96	59.74	85.75	82.48
March	96.78	216.73	118.64	90.73	94.35	97.99
April	164.60	176.31	138.25	147.56	126.76	161.85
May	89.68	115.19	100.83	70.53	86.20	89.88
June	74.95	91.12	85.19	72.28	61.19	59.94
July	47.17	54.24	42.23	22.18	35.55	40.37
August	42.35	60.28	41.55	28.87	30.56	37.39
September	73.91	68.41	62.87	51.80	58.23	62.01
October	144.43	160.03	139.38	153.00	101.85	110.57
November	259.11	196.39	147.42	174.88	187.87	174.24
December	116.76	243.08	114.95	103.87	118.72	125.64

Further, Table 3.5 reveals that the variability measured in terms of coefficient of variation is very high for the individual months, particularly for the months having a small proportion of the annual total. Similarly, the months having higher proportion of the annual total have a lower coefficient of variations. The monthly shares also shows as much variability as in the absolute rainfalls. Moreover, the variability in the former is strongly and positively correlated (r = +0.89) with the latter.

Though variation in seasonal or monthly rainfall is significantly higher than the variation in annual total rainfall, it should not lead one to suggest that annual totals have no significant or that they have no relation with seasonal shortfalls. Whereas marginal shortfalls in monthly rainfall may be made good, prolonged and significant shortfalls are bound to be reflected in the annual total. The Seasonal Negative Deviation Index (SNDI) for each year and district for the period 1971 to 1993 is derived as follows:

$$SNDI = \frac{\sum_{i=1}^{n} XiYi}{n.100}$$

Where Xi = below normal deviations of rainfall in the ith month

Yi = normal percentage share of the ith month, and

n = number of months (12)

This is a measure of average deficiency in rainfall weighted by the normal share of each month, not netted for excess of rainfall in other months. Last column of Table 3.6 revealed that the coefficient of correlation between SNDI and annual total rainfall of respective year for all the districts under study. All the correlation coefficients are negative and statistically high except in heavy rainfall district of Amritsar and Gurdaspur. The corresponding value of the correlation coefficient were –0.632 and –0.641. This would mean that higher the seasonal deficiency in any year (not counting positive deviations from the normal) lowest would be the annual total.

Table 3.6: Coefficient of correlation (r) between monthly and annual rainfalls at selected stations in Punjab

Station(s)	Coefficient of Correlation (r) between Annual and Monthly Rainfall of												r-SNDI[$]
	Jan	Feb.	March	April	May	June	July	August	September	October	November	December	
Amritsar	0.055	0.289	0.173	0.124	0.114	0.321	0.574	0.811	0.628	0.183	0.180	0.013	–0.632‡
Bathinda	0.274	0.015	0.364*	0.042	0.238	0.302*	0.365*	0.868‡	0.594‡	0.314*	0.342*	0.463†	–0.749‡
Ferozepur	0.042	0.467†	0.151	0.375*	0.289	0.287	0.361*	0.722‡	0.631‡	0.089	0.184	0.113	–0.787‡
Gurdaspur	–0.008	0.019	0.035	0.173	0.315*	0.490	0.573‡	0.651‡	0.549†	0.019	0.127	0.342*	–0.641‡
Ludhiana	0.563‡	0.134	0.022	0.487†	0.509†	0.397†	0.148	0.709‡	0.353*	0.270	0.251	0.225	–0.798‡
Patiala	0.504†	0.214	0.033	0.466†	0.535‡	0.215	0.856‡	0.610‡	0.575‡	0.243	0.183	0.181	–0.803‡

‡ implies significance at 0.01 level
† implies significance at 0.05 level
* implies significance at 0.10 level
$ = r between SNDI and Annual Total

This relation is quite strong and it is unlikely that annual total is normal or above normal inspite of high deficiency of rainfall in different months. While timeliness of rainfall is important, its importance as a dimension separate from annual total tends to be often exaggerated. Even in drought-prone districts we find that the two go together fairly steadily.

We may finally raise the question of whether rainfall during some early months could be used to make atleast an approximate forecast of rainfall of the annual total or of subsequent months during the year concerned. We computed correlation[10] between rainfall of different months with each other and presented in Annex Table 3.11 to 3.16. The correlation coefficients between the monthly rainfall and the annual totals are presented in Table 3.6. A close look at Table 3.6 reveals that strongest correlations coefficient was observed between annual total rains with rainfall of August for all the rain gauge stations (except in Patiala where the strongest correlation was for the month of July). Likewise correlation coefficient between the annual total rains and September rains were also statistically significant. Besides a few other statistically significant correlations emerged, but there magnitude was not enough and predictions would seems hazardous. Hence one may broadly say that if rains fail in August and September in any district, one may predict that is not likely to be a good rainfall year in that particular district.

The correlation coefficients between individual months (Annex Table 3.11 to 3.16) was much less strong, making it difficult to forecast rainfall of subsequent individual months from a given month's rainfall. Statistically, significant correlation coefficient was observed between April-May rainfall in Gurdaspur, Patiala and Ludhiana, July-August rainfall for Patiala, Ludhiana and Amritsar and August–September rainfall for Gurdaspur, Ferozepur and Bathinda district. Interestingly, there were a number of cases of negative and statistically significant correlation, the strongest observed being –0.435 in Amritsar district between rainfall of January and March. If rains, are below normal in January, the chances are that they would be above normal in March in this district. Likewise, correlation coefficient between the rains of March and December was observed at –0.427 in case of

Gurdaspur district. This implies that if rains are below normal in the month of March, it is likely that rains would be above normal in December. These are cases of shortfall in one month but are made good in another. It may be noted that the negative correlation are not quite strong and do not dominate the picture.

REFERENCES

1. N.R. Kothare, "Rainfall in India," *Reserve Bank of India—Staff Occasional Papers*, Vol. 2, No. 2, December 1977.
2. G.O.I. *Report of the National Commission of Agriculture, Climate and Agriculture*, Vol. 4, Ministry of Agriculture and Irrigation, New Delhi, 1976.
3. For detail please see M.G. Kendall, *Advanced Theory of Statics*, Vol. 2, 1958.
4. Louis Bean, "A New Approach to Statistical Forecasting of Next Year's Weather and Crops," *Agricultural Situation in India*, Vol. 23, No. 12, March 1969.
5. Dr. M.S. Swaminathan was reported to have informed this of. Please see *The Hindu*, October 8, 1980.
6. Ferozepur claimed much more canal irrigation water as the farmers have shown more area under orchards than actually.
7. An analysis of weekly rainfall data is not attempted here. For the purpose of our analysis such minuteness was not worthwhile. These data are not published, and may be locally available for given areas at which level they are easy to handle and also worthwhile to analyse. In the process of averaging across rain gauge stations, week-by-week fluctuations get lost at the district level. Sowing weeks differs widely between crops and even areas within a district. The hypothesis that crops need week-by-week regularity in rain can be tested at the level of experiment station rather than a district. However, nature is hardly a paradise of such regularity and crops have some ability to survive short spell of drought.
8. Like annual rainfall, no definite trend has been noticed in rainfall during different seasons. The estimates of number of turning points and its associated Z-statistics are reported in Annex Table 3.2
9. The application of Z-statistic (Kendall, 1958) did not exhibiting any definite trend in the rainfalls during different months. The estimates of number of turning points and its associated Z-statistics are reported in Annex Table 3.3.
10. The Correlation of Rainfall between months for Meteorological Subdivision are already Computed. Please see K.N. Rao, S. Jaijanthi and V.K. Bhargav, *Indian Monsoon Correlation—Part I, Monthly Intercorrelation for all the Meteorological Subdivisions of India*, Meteorological Monogarph No. 4, Climatory, I.M.D., Poona 1972. However we are computing the monthly correlation at district level i.e., at a more disaggregated level.

Annex Table 3.1
Estimates of linear time trend of average annual rainfall at selected stations in Punjab: 1971 through 1993

Station(s)	Pure Intercept	Estimated Coefficients on Time Variable	$(\bar{R})^2$
Amritsar	55.314	+0.989 ns (0.935)	0.058
Bathinda	35.365	–0.055 ns (0.253)	0.003
Ferozepur	30.678	+1.026 ns (0.783)	0.167
Gurdaspur	101.213	+0.039 ns (0.214)	0.001
Ludhiana	54.549	+0.846 ns (0.675)	0.116
Patiala	46.831	+1.753 ns (1.375)	0.173

Figures in parentheses are standard error of the respective coefficient.
ns implies not significant

Annex Table 3.2
Estimates of turning points and corresponding Z-statistics for seasonal rainfall at selected stations in Punjab: 1971 through 1993

Station(s)	Winter Rains (WSR)		Pre-Monsoon (PMR)		South-West Monsoon(SWR)		North-East Monsoon(NER)	
	NTP	Z-value	NTP	Z-value	NTP	Z-value	NTP	Z-value
Amritsar	13	0.56	12	—	13	0.56	15	1.67
Bathinda	11	0.56	12	—	12	—	13	0.56
Ferozepur	13	0.56	15	1.67	11	0.56	12	—
Gurdaspur	15	1.67	14	1.11	13	0.56	15	1.67
Ludhiana	13	0.56	19	3.89	12	—	11	0.56
Patiala	13	0.56	12	—	13	0.56	13	0.56

NTP implies number of turning points.

Annex Table 3.3
Number of turning points and Z-statistics of monthly rainfall at selected stations in Punjab: 1971 through 1993

Month(s)	Amritsar	Bathinda	Ferozpur	Gurdaspur	Ludhiana	Patiala
			Number of Turning Points			
January	12	12	13	15	13	13
February	13	11	11	11	13	15
March	9	10	10	10	10	10
April	12	12	13	13	15	12
May	15	11	15	14	18	13
June	15	12	12	15	14	13
July	12	13	9	11	9	11
August	12	10	14	11	10	12
September	13	11	11	13	0	14
October	12	16	16	10	16	15
November	10	11	12	9	11	12
December	11	9	13	11	11	11
			Estimated Value of Z-statistics			
January			0.56	1.67	0.56	0.56
February	0.56	0.56	0.56	0.56	0.56	1.67
March	1.67	1.11	1.11	1.11	1.11	1.11
April			0.56	0.56	1.67	
May	1.67	0.56	1.67	1.11	3.34	0.56
June	1.67			1.67	1.11	0.56
July		0.56	1.67	0.56	1.67	0.56
August		1.11	1.11	0.56	1.11	
September	0.56	0.56	0.56	0.56		1.11
October		2.22	2.22	1.17	2.22	1.67
November	1.11	0.56		1.67	0.56	
December	0.56	1.67	0.56	0.56	0.56	0.56

Annex Table 3.4
Probability of occurrence (as per cent to total years) of seasonal rainfall in different ranges from normal (average) rainfall at selected stations in Punjab: 1971 through 1993

Station(s)	Ranges of deciles from normal (Per cent)						
	Failure/ Below Normal			Normal		Excess/Above Normal	
	Below –50	–30 to –50	–10 to –30	–10 to +10	+10 to +30	+30 to +50	Above +50
Winter Rains (WSR)							
Amritsar	10	15	5	45	5	5	15
Bathinda	30	15	15	0	10	5	25
Ferozepur	20	20	0	10	15	20	15
Gurdaspur	20	0	15	20	25	5	15
Ludhiana	20	10	10	25	15	0	20
Patiala	25	20	0	10	15	20	15
Hot weather Rains (PMR)							
Amritsar	20	20	30	0	5	0	25
Bathinda	35	10	15	5	5	10	20
Ferozepur	20	20	25	10	5	0	20
Gurdaspur	20	15	30	15	5	0	15
Ludhiana	25	15	15	25	0	0	20
Patiala	30	5	15	25	10	5	10
South-West Rains (SWR)							
Amritsar	5	25	25	15	10	5	15
Bathinda	10	10	20	20	25	5	10
Ferozepur	5	15	40	5	20	5	10
Gurdaspur	0	15	10	60	5	5	5
Ludhiana	5	15	10	30	30	10	0
Patiala	5	15	5	40	25	0	10
North-West Rains (NWR)							
Amritsar	15	25	30	20	0	0	10
Bathinda	30	10	20	10	10	5	15
Ferozepur	35	5	25	10	5	0	20
Gurdaspur	35	20	20	0	10	0	15
Ludhiana	30	15	5	20	15	0	15
Patiala	40	5	25	5	5	0	20

Annex Table 3.5
Probability of occurrence (as per cent to total years) of rainfall in different ranges from normal (average) monthly rainfall in Amritsar: 1971 through 1993

Station(s)	Ranges of decilies from Normal (Per cent)						
	Failure/ Below Normal			Normal		Excess/Above Normal	
	Below −50	−30 to −50	−10 to −30	−10 to +10	+10 to +30	+30 to +50	Above +50
January	30	5	10	5	0	30	20
Feburary	20	0	40	15	0	5	20
March	35	5	25	5	15	0	15
April	55	0	20	5	20	0	0
May	25	40	5	5	0	0	25
June	20	25	15	5	15	0	20
July	10	5	40	15	15	5	10
August	15	15	30	10	0	20	10
September	40	5	20	15	0	0	20
October	35	20	10	10	0	5	20
November	60	0	5	5	15	0	10
December	35	5	25	10	10	0	15

Annex Table 3.6
Probability of occurrence (as per cent to total years) of rainfall in different ranges from normal (average) monthly rainfall in Bathinda: 1971 through 1993

Station(s)	Ranges of decilies from normal (Per cent)						
	Failure/ Below Normal			Normal		Excess/Above Normal	
	Below −50	−30 to −50	−10 to −30	−10 to +10	+10 to +30	+30 to +50	Above +50
January	40	10	5	5	0	5	35
Feburary	45	0	10	15	5	5	20
March	50	5	5	15	0	10	15
April	45	15	0	0	5	0	35
May	45	10	10	5	10	0	20
June	30	10	10	15	5	5	25
July	20	10	15	20	15	5	15
August	25	15	20	5	5	5	25
September	25	20	10	5	0	20	20
October	45	5	0	25	5	5	15
November	75	0	0	0	0	10	15
December	55	5	10	5	0	10	15

Annex Table 3.7

Probability of occurrence (as per cent to total years) of rainfall in different ranges from normal (Average) monthly rainfall in Ferozepur : 1971 through 1993

Station(s)	Ranges of decilies from normal (per cent)						
	Failure/ Below Normal			Normal		Excess/Above Normal	
	Below −50	−30 to −50	−10 to −30	−10 to +10	+10 to +30	+30 to +50	Above +50
January	50	0	0	0	5	0	45
Feburary	20	15	10	30	5	10	10
March	35	5	25	5	5	15	10
April	55	5	20	0	0	0	20
May	40	20	5	0	5	5	25
June	25	10	5	25	10	0	25
July	10	5	45	20	5	10	5
August	20	15	25	10	15	5	10
September	35	0	0	30	10	5	20
October	45	25	0	0	0	5	25
November	45	20	5	0	0	10	20
December	45	5	10	20	5	0	5

Annex Table 3.8

Probability of occurrence (as per cent to total years) of rainfall in different ranges from normal (average) monthly rainfall in Gurdaspur: 1971 through 1993

Station(s)	Ranges of decilies from normal (Per cent)						
	Failure/ Below Normal			Normal		Excess/Above Normal	
	Below −50	−30 to −50	−10 to −30	−10 to +10	+10 to +30	+30 to +50	Above +50
January	35	5	0	5	20	5	30
Feburary	15	15	10	20	5	15	20
March	30	10	30	5	5	5	15
April	40	5	40	0	0	5	10
May	25	5	25	5	15	0	25
June	25	25	5	5	10	10	20
July	0	15	15	40	20	0	10
August	10	10	10	45	15	0	10
September	20	10	5	15	20	15	15
October	30	35	0	20	5	10	0
November	55	0	15	5	0	0	25
December	40	0	25	0	5	0	30

Annex Table 3.9
Probability of occurrence (as per cent to total years) of rainfall in different ranges from normal (average) monthly rainfall in Ludhiana: 1971 through 1993

	Ranges of decilies from Normal (per cent)						
	Failure/ Below Normal			Normal		Excess/Above Normal	
Station(s)	Below –50	–30 to –50	–10 to –30	–10 to +10	+10 to +30	+30 to +50	Above +50
January	40	10	0	0	5	25	20
Feburary	25	25	0	20	5	5	20
March	15	0	10	15	20	10	30
April	50	0	0	25	5	5	15
May	35	20	10	10	0	0	25
June	15	15	25	5	5	10	25
July	5	20	5	30	25	10	5
August	15	5	35	10	5	15	15
September	25	10	10	5	10	30	10
October	45	0	0	0	0	25	30
November	70	0	20	0	0	0	10
December	45	10	20	5	0	0	20

Annex Table 3.10
Probability of occurrence (as per cent to total years) of rainfall in different ranges from normal (average) monthly rainfall in Patiala: 1971 through 1993

	Ranges of decilies from normal (Per cent)						
	Failure/ Below Normal			Normal		Excess/Above Normal	
Station(s)	Below –50	–30 to –50	–10 to –30	–10 to +10	+10 to +30	+30 to +50	Above +50
January	40	5	5	0	30	5	15
Feburary	15	20	10	20	20	0	15
March	40	5	5	0	0	5	45
April	45	15	25	0	0	0	15
May	45	0	20	5	5	0	25
June	10	15	15	20	10	10	20
July	15	5	10	30	10	15	15
August	10	0	15	45	10	15	5
September	20	15	5	25	5	0	30
October	45	10	5	0	0	25	15
November	55	0	5	0	25	0	15
December	45	15	10	5	0	5	20

Annex Table 3.11
Coefficient of Correlation (r) between monthly rainfalls in Amritsar

Month(s)	January	February	March	April	May	June	July	August	September	October	November	December
January	1.000											
Feburary	–0.003											
March	–0.326*	0.453†										
April	0.634‡	0.114	–0.085									
May	0.567‡	0.087	–0.157	0.727‡								
June	–0.127	0.028	0.134	–0.243	–0.049							
July	–0.066	0.164	–0.049	–0.071	–0.165	0.048						
August	0.087	0.014	–0.139	0.042	–0.014	0.209	0.392*					
September	–0.280	0.412	0.148	–0.092	–0.045	0.274	0.163	0.427				
October	–0.088	–0.217	–0.053	0.217	0.105	–0.072	0.076	0.174	–0.199			
November	–0.038	0.080	0.063	–0.192	–0.200	–0.151	0.040	–0.449	–0.180	–0.136		
December	–0.435†	–0.236	0.339†	–0.178	0.184	–0.191	0.064	–0.046	–0.396†	0.319*	–0.015	

‡ implies significance at 0.01 level
† implies significance at 0.05 level
* implies significance at 0.10 level

Annex Table 3.12

Coefficient of Correlation (r) between monthly rainfalls in Bathinda

Month(s)	January	February	March	April	May	June	July	August	September	October	November	December
January	1.000											
Feburary	0.178											
March	–0.239	–0.026										
April	0.253	–0.122	0.066									
May	–0.018	0.046	–0.094	–0.042								
June	–0.017	–0.110	–0.382†	–0.330*	0.069							
July	–0.328	–0.233	–0.249	–0.221	–0.005	0.025						
August	0.370*	0.009	–0.362*	0.137	0.219	0.346*	–0.015					
September	0.396†	0.128	–0.398†	0.073	–0.051	0.156	–0.143	0.603‡				
October	0.283	–0.208	–0.125	0.809‡	–0.002	–0.193	0.139	0.068	0.268			
November	–0.218	0.049	0.044	–0.068	–0.268	–0.319*	–0.122	–0.166	–0.285	–0.298		
December	–0.194	–0.190	0.884	–0.076	0.093	–0.346	–0.418	–0.338*	–0.369*	–0.195	–0.173	

‡ implies significance at 0.01 level
† implies significance at 0.05 level
* implies significance at 0.10 level

Annex Table 3.13
Coefficient of Correlation (r) between monthly rainfalls in Ferozepur

Month(s)	January	February	March	April	May	June	July	August	September	October	November	December
January	1.000											
Feburary	0.017											
March	–0.066	0.130										
April	0.477†	–0.055	0.058									
May	–0.062	–0.026	–0.154	0.220								
June	–0.022	–0.062	–0.359	0.174	0.348*							
July	0.060	–0.066	0.015	0.036	–0.121	–0.169						
August	0.125	0.251	0.070	0.072	–0.056	–0.125	–0.118					
September	0.001	0.544‡	0.165	0.056	0.085	0.271	–0.100	0.386†				
October	0.009	0.162	–0.037	0.118	0.131	0.467†	0.536‡	–0.268	0.089			
November	–0.218	–0.046	0.815‡	–0.132	–0.209	–0.332*	0.310	–0.109	–0.035	0.083		
December	–0.176	–0.175	–0.180	–0.070	0.543‡	0.284	0.186	0.073	–0.276	0.126	–0.024	

‡ implies significance at 0.01 level
† implies significance at 0.05 level
* implies significance at 0.10 level

Annex Table 3.14
Coefficient of Correlation (r) between monthly rainfalls in Gurdaspur

Month(s)	January	February	March	April	May	June	July	August	September	October	November	December
January	1.000											
Feburary	0.151											
March	–0.105	0.362*										
April	0.445†	0.212	0.048									
May	0.239	0.092	0.029	0.634‡								
June	–0.172	0.100	–0.301*	–0.152	0.377*							
July	0.117	–0.167	–0.154	–0.160	–0.155	0.126						
August	–0.104	–0.254	–0.264	–0.083	–0.074	0.161	0.297					
September	–0.416	0.133	–0.030	0.027	0.372*	0.638‡	–0.152	0.362*				
October	–0.186	–0.291	–0.133	–0.037	–0.142	–0.107	0.166	0.006	–0.158			
November	0.173	0.056	–0.170	–0.122	–0.313*	0.093	0.532‡	–0.186	–0.127	–0.099		
December	–0.275	0.348	–0.427	–0.097	–0.064	–0.108	0.383	0.111	–0.115	0.140	0.137	

‡ implies significance at 0.01 level
† implies significance at 0.05 level
* implies significance at 0.10 level

Annex Table 3.15
Coefficient of correlation (r) between monthly rainfalls in Ludhiana

Month(s)	January	February	March	April	May	June	July	August	September	October	November	December
January	1.000											
Feburary	–0.046											
March	0.068	0.041										
April	0.702‡	–0.136	–0.091									
May	0.448†	–0.254	–0.231	0.568‡								
June	0.107	–0.054	–0.442†	–0.158	0.301*							
July	–0.036	0.111	–0.197	0.233	0.118	–0.179						
August	0.549	–0.209	–0.245	0.379	0.269	–0.385*	0.341*					
September	0.129	–0.238	0.123	0.122	0.085	0.189	–0.255	0.129				
October	0.123	–0.283	0.138	0.143	0.064	–0.254	0.166	0.146	–0.211			
November	–0.158	0.061	0.128	–0.155	–0.234	–0.218	–0.123	–0.330*	–0.321*	–0.247		
December	–0.074	0.425†	0.657‡	–0.183	–0.067	–0.210	–0.172	–0.010	0.066	0.346*	0.177	

‡ implies significance at 0.01 level
† implies significance at 0.05 level
* implies significance at 0.10 level

Annex Table 3.16
Coefficient of correlation (r) between monthly rainfalls in Patiala

Month(s)	January	February	March	April	May	June	July	August	September	October	November	December
January	1.000											
Feburary	–0.013											
March	0.171	0.027										
April	0.724‡	–0.117	0.173									
May	0.473†	0.065	–0.168	0.677‡								
June	0.080	0.073	–0.500†	–0.280	–0.047							
July	0.182	–0.149	–0.234	0.248	0.352*	0.124						
August	0.258	–0.268	–0.237	0.230	0.283	0.274	0.409†					
September	0.227	–0.293	0.180	0.046	0.289	–0.054	0.431†	0.169				
October	0.006	–0.358	0.065	–0.083	–0.242	0.280	0.306	–0.137	0.121			
November	–0.178	–0.161	0.259	–0.121	–0.177	–0.179	0.326*	–0.266	0.102	0.273		
December	–0.286	–0.365*	0.288	–0.145	–0.231	–0.036	0.227	–0.222	0.153	0.429†	0.788‡	

‡ implies significance at 0.01 level
† implies significance at 0.05 level
* implies significance at 0.10 level

Annex Table 3.17

Estimated value of β_1 and β_2 of annual and seasonal rainfalls at selected stations in Punjab: 1971 through 1993

Season(s)	Amritsar		Bathinda		Ferozepur		Gurdaspur		Ludhiana		Patiala	
	β_1	β_2	β_1	β_2	β_1	β_2	β_1	β_2	β_1	β_2	β_1	β_2
Annual rains	0.92	0.18	0.40	21.40	0.02	0.90	0.09	0.19	0.01	0.14	0.01	0.22
Winter Rains	0.00	0.17	0.03	0.14	0.00	0.13	0.00	0.13	0.00	0.14	0.05	0.12
Pre-monsoon Rains	0.17	0.33	0.07	0.19	0.16	0.06	0.22	0.35	0.18	0.33	0.41	0.59
South-west Rains	0.02	0.16	0.00	0.19	0.06	0.23	0.10	0.27	0.01	0.12	1.06	0.19
North-east Rains	0.32	0.43	0.13	0.26	0.05	0.14	0.15	0.22	0.10	0.23	0.22	0.38

Annex Table 3.18

Estimated value of β_1 and β_2 of monthly rainfalls at selected stations in Punjab: 1971 through 1993

Month(s)	Amritsar		Bathinda		Ferozepur		Gurdaspur		Ludhiana		Patiala	
	β_1	β_2	β_1	β_2	β_1	β_2	β_1	β_2	β_1	β_2	β_1	β_2
January	0.00	0.13	0.04	0.14	0.02	0.12	0.00	0.09	0.07	0.22	0.15	0.31
Feburary	0.02	0.20	0.10	0.20	0.10	0.27	0.00	0.12	0.04	0.14	0.09	0.24
March	0.18	0.27	0.47	0.62	0.32	0.52	0.15	0.30	0.22	0.40	0.02	0.11
April	0.46	0.60	0.59	0.71	0.40	0.59	0.65	0.76	0.47	0.61	0.76	0.84
May	0.13	0.25	0.14	0.25	0.17	0.06	0.04	0.15	0.06	0.16	0.09	0.20
June	0.06	1.16	0.08	0.23	0.08	0.20	0.16	0.28	0.02	0.14	0.01	0.18
July	0.10	0.30	0.03	0.15	0.52	0.68	0.03	0.08	0.02	0.14	0.02	0.14
August	0.15	0.37	0.03	0.13	0.39	0.57	0.05	0.21	0.00	0.11	0.01	0.20
September	0.04	0.16	0.01	0.11	0.03	0.12	0.04	0.19	0.00	0.11	0.02	0.16
October	0.63	0.75	0.43	0.59	0.25	0.40	0.66	0.77	0.01	0.10	0.11	0.23
November	0.15	0.22	0.32	0.45	0.09	0.15	0.52	0.67	0.16	0.32	0.62	0.74
December	0.31	0.45	0.46	0.60	0.22	0.33	0.20	0.35	0.12	0.20	0.27	0.41

CHAPTER IV

GROWTH AND FLUCTUATIONS IN CROP PRODUCTION

Reliable estimation of trends in agricultural output is more than a matter of mere academic interest. Their use in predicting the future may hardly be dependable; but as a measure of performance during a period, they have considerable significance to policy makers. Aggregate[1] growth rates, however, in agriculture offer no clue to the disparate performance of agriculture in different regions in India. Estimations of growth rates by crops for different regions across states followed by explanations on why they differ are crucial for an insight into the agricultural development process. But the question as to how different regions have responded to the changing economic conditions of the country and how Indian agriculture has been doing in the last few decades, have been the centre of debate among the various economists.[2] The question has got obvious political and other undertones and a great number of very loose statements was made by politicians, journalists and vocal sections of the public as to how Indian agriculture had passed through a period of green revolution and recording remarkable rates of growth or how the same agriculture was stagnating under the weight of conditions described as 'semi feudal', etc. Most of the statements were of course made without basing them on any adequate empirical data or any careful statistical analysis of the same.

The normal statistical procedure to obtain a measure of growth of output over the period of a series is to postulate a hypothetical function which would adequately describe the series of outputs overtime and to estimate its parameters which would offer a measure of growth of output over the period. In strict logic, the function is chosen on *a priori* grounds, such as, a theory or a hypothesis underlying the set of observations. The output of agriculture in a year is affected and determined by innumerable factors and circumstances and it does not seem possible to theorize the growth of agricultural output over a period. The idea that a better understanding of the crop-weather-input relationship at the micro level will enable such a theorizing is only a wishful thinking. But this need not prevent certain loosely defined or broad *a priori* considerations to govern the choice of a functional form. For instance, we may ask: Would the change in agricultural output in a given year be independent of the size of output in the preceding year or would it depend on it? On purely *a priori* grounds, it seems reasonable to suppose that it would so depend. We should make it clear that this is not necessarily true of change in everything and that, even in relation to changes in agricultural output, one may not agree with this *a priori* reasoning. Nevertheless, we suggest that the postulate that the change in agricultural output in a year would depends upon the output in the preceding year is reasonable and that we should accept it. In short, what we decide on a *priori* consideration is that of the choice of the function is limited to a simple functions of time (t), we should suppose that if (O) is output, then not (O) but log (O) should be supposed to follow such functions over time.[3] We repeat that the choice is made on *a priori* considerations but that we consider these considerations reasonable.

It assumes that the profile over time of the output can be decomposed into two neat components (i) a systematic component representable by a smooth function of time, and (ii) a small error component that behaves like a random variable having a known and and a well behaved probability distribution. But what usually catches one's eye in an area of econometric research is the glittering display of alternative functional forms. They thrive best in situations of plentiful data for statistical play or where the researchers can employ *a priori* theoretical reasoning to pin down the forms of relationships. Since we don't have any theoretical justification for the choice of particular functional form, we have to fall back on an entirely empirical approach. In the situation, we are in search of a satisfactory approximation of the true function so that the residual left are fairly random in nature and adequately satisfy the known and well behaved probability distribution, namely, independently and identically distributed with mean equal to zero, that is,

$$E(u_i) = 0 \text{ for all } i \text{ and}$$

$$E(UiUj)\begin{cases} 0 \text{ for all } i \neq j \\ \sigma^2 \text{ for all } i = j \end{cases}$$

where i and j = 1,2,3. . .n

Growth Curves

Suppose we have time series data Y(t) on a characteristic Y. It is well known that the change in Y per unit of time at (t) is denoted by :

$$\frac{dYt}{dt} \text{ or } Y(t+1) - Y(t)$$

according as (t) varies continuously or takes only discrete values. Since we are dealing with a time series, we say that there is growth or stagnancy or decay (same as retardation) in Y at (t) according as dY(t)/dt is positive or equal or negative respectively.

The measure which is of importance to us is the rate of change in Y per unit of time at (t) usually termed as 'Growth Rates' at (t) denoted by G(t) and defined as:

$$G(t) = \frac{1}{Y(t)} \frac{dYt}{dt}$$

If the observation Y(t) is on a continuous time scale, and

$$G(t) = \frac{Y(t+1) - Y(t)}{Y(t)}$$

If the observation Y(t) is on a discrete scale.

The smooth curve passing through the points of the graph of Y(t) against (t) is called the growth curve. From the growth curve one can easily obtain the quantitative characterisation of the path of growth of the characteristic under consideration. The phenomenon of G(t) is rising (falling) over time is called acceleration (deceleration) in growth of Y's curve.

In considering the dynamics of production as a function of time (t), one should initially anaylse the applicability of such very simple functions as:[4]

$$\log Y = a + bt \qquad \text{(SLM)} \qquad \text{I}$$
$$\log Y = (a + bt)^{-1} \qquad \text{(LIM)} \qquad \text{II}$$
$$\log Y = a + bt + ct^2 \qquad \text{(LQM)} \qquad \text{III}$$

There is, however, said to be a problem of multicollinearity posed by LQM functional form because of high correlation between (t) and (t^2). The correlation between (t) and (t^2) for any span of two years or more but without gaps, can be shown to be bounded below by $(15/16)^{1/2}$, a very high value. As such, the problem of multicollinearity raises its ugly head[5] particularly when the observations are few.

In order to avoid the high correlation between (t) and (t^2), we have measured time (t) not from the beginning of the first year (calling the first year as 1) but from the mid-point of the period. Thus, we have transformed (t) as follow:

$$t' = t - (n + 1)/2$$

where n is the number of years for which we have the time series data. Correlation between (t') and $(t')^2$ is zero, that is, these are orthogonal. The modified functional form of III will be:

$$\log Y = a + bt' + ct'^2 \quad \text{(MQM)} \qquad \ldots \qquad \text{IV}$$

All these functional forms were fitted through the method of ordinary least (OLS) squares.

Recent years have brought some relief from the monotony of linear and compound growth rates. The trend equations underlying these rates are unsuitable when one has ground to suppose that growth may be proceeding at accelerated or decelerated rates rather than at constant rate. For examplĕ, it is plausible that the 'green revolution' in Indian agriculture has imparted a fresh momentum to growth which would suggest an acceleration in growth rates; equally plausible, the declining importance since the sixties of the contribution to growth by expansion in areas sown could have brought about deceleration. This is the kind of situation requiring relatively flexible functional forms for the trend equation which could leave the question of acceleration or deceleration open for the data in hand to decline. Therefore, we felt need to introduce the two more mathematical growth curves, namely, Gompertz and Logistic.

In the present study, following logistic curve (LCM) has been fitted:

$$1/Y = kab^t \quad \ldots \quad V$$

The equation of the Gompertz (GCM) curve is:

$$Y = Ka^{b^t} \quad .. \quad VI$$

which may be put in logarithmic form as:

$$\log Y = \log K (\log a)b^t \quad \ldots \quad VII$$

The computation of the logistic and the Gompertz curves were through the method of selected points. The logistic resembles the Gompertz and in that sense, these two are competing curves.

CHOICE OF GROWTH CURVES

In order to find out a satisfactory approximation of the true function, we begin with the examination of the residuals corresponding to each fitted function. Now, there can be more than one functional form which satisfy the assumption about the behaviour of the residuals. In such cases, we shall choose the one with the highest R^2 corrected for degree of freedom, which permits the comparison of the explanatory powers of the two functions with different number of explanatory variables. We also introduced a measure known as 'C' test; which can be defined as:

$$C = \frac{(P - Q)}{(n - 1)}$$

where P is the number of pairs of adjustment residuals of the same sign, and Q is the number of pairs of adjacent residuals of the opposite sign. Obviously, (P + Q) = (n – 1) and this 'C' is an indicator of first order auto correlation coefficient of residuals. This coefficient 'C' helps us to identify the best fitted curve where the different fitted trend curves give more or less the same value of $\overline{R}^2$ value.[6] Thus, out of two alternative forms of curves with more or less the same $\overline{R}^2$ value, one must choose that form which bears the lower value of 'C', indicates thereby the little fluctuations in the observed pattern of the variable to be explained.

GROWTH ANALYSIS

We have taken up the growth analysis at the disaggregated level, that is, of the selected crops, namely, wheat, paddy, sugarcane, cotton desi, cotton american and rapeseed and mustard, for the selected districts. The time reference of the analysis is twenty one years spanning the period 1971-72 through 1991-92. The dispersion (S.D.); the coefficient of variation (C.V.) and the algebraic sum of the difference between the actual values and the theoretical ones are presented in Annex Table 4.1. These results confirmed the theoretical and empirical presumptions made earlier concerning the unsuitability of the semi-logarithmic exponential function (SLM) which is due to the existence of an upper/lower symptoms boundary of the growth of crop production and the law of continuously declining/increasing annual rates of growth. These results are discussed separately as follows for each crop:

WHEAT OUTPUT

The computation of the parameters of the growth curves (I) to (IV) through ordinary least square (OLS) method and of (V) and (VI) through the selected ends points yield the theoretical growth curves as presented in Table 4.1.

The estimated value of both the measure of the goodness of fit, namely C-value $\bar{R}^2$ respectively are presented in Annex

Table 4.1: Estimated growth curves for wheat output

Model(s)	Estimated value of the coefficients for			
	K	a/log a	b	c
		Ludhiana		
SLM		6.786	0.037‡ (6.786)	
LIM		0.151	–0.002‡ (6.382)	
LQM		6.645	+0.075‡ (3.829)	–0.002‡ (2.123)
MQM		6.693	+0.035‡ (7.754)	–0.002‡ 2.123)
LCM	389.961	–0.252	+0.038	
GCM	2.703	+0.112	+1.058	
		Gurdaspur		
SLM		5.659	+0.058‡ (10.724)	
LIM		0.179	–0.002‡ (9.385)	
LQM		5.283	+0.079‡ (3.498)*	–0.001 ns (0.986)
MQM		5.967	+0.057‡ (10.716)	–0.001 ns (0.986)
LCM	725.405	–0.450	+0.053	
GCM	1.946	+0.479	+1.032	

The computations of first four growth curves are in natural (base e) logarithmic while that of last two in base 10.

Figures in parentheses are the 't' value of the respective parameters.

‡ implies significance at 0.01 level.

† implies significance at 0.05 level.

* implies significance at 0.10 level.

ns not significant

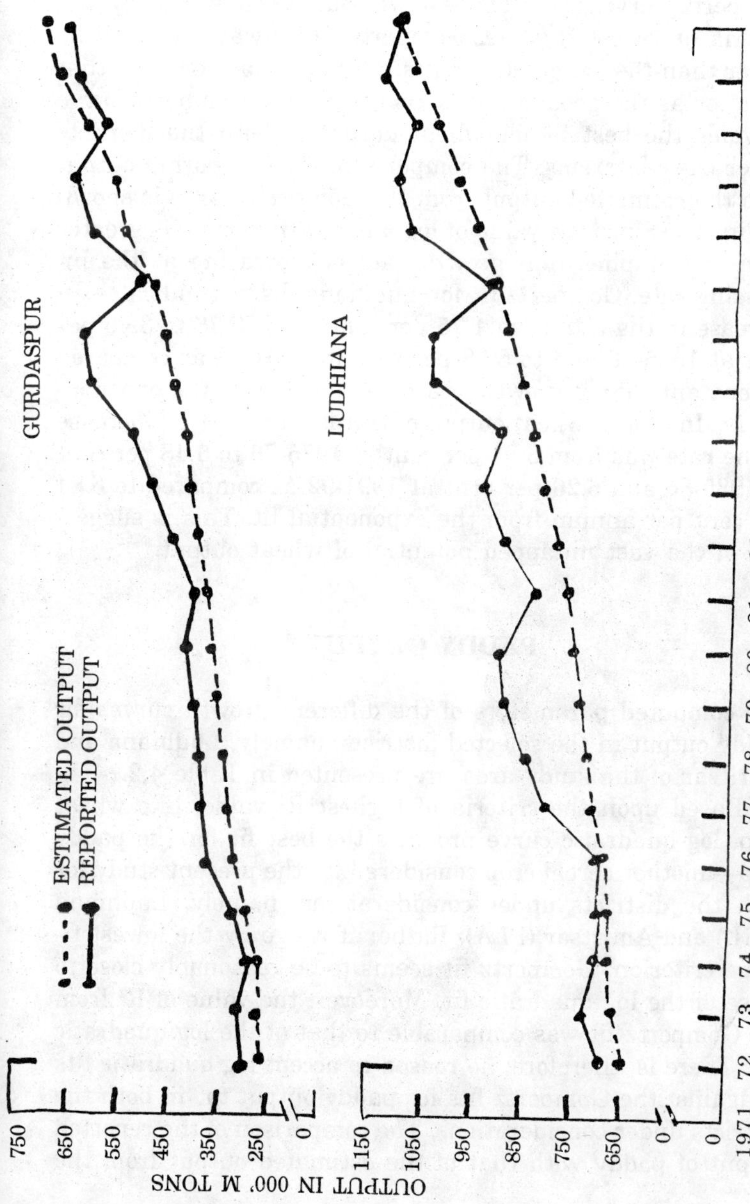

Fig. 4.1: Comparison of estimated and reported output of wheat: 1971–72 through 1991–92

Tables 4.2 and 4.3. According to highest $(\overline{R})^2$ criteria, log-quadratic fits seems to be the best fit for wheat output of Ludhiana district (WHL) while for the Gurdaspur (WLG), Gompertz curve gave the best fit. But when we go by the criteria of lowest C-value, Gompertz fits always seem to be better than the log quadratic fits. Thus we may conclude that in so for as the goodness of fit is concerned, Gompertz curve provides the best fit for wheat output of both the districts under considerations. The comparison of the reported output with the estimated output from the Gompertz curve is shown in Fig. 4.1. Since the value of log a is positive and b is greater than one, implies that growth rate is increasing at the increasing rate. Gompertz fit for Ludhiana district indicates an increase in the rate from 1.75 per cent in 1975-76 to 3.46 per cent at 1985-86 and to 5.58 per cent in 1991-92 as compared to constant rate of growth of 3.67 per cent from the exponential fit. In case of wheat output of Gurdaspur district, increase in the rate was from 3.81 per cent at 1975-76 to 5.48 per cent at 1985-86 and 6.26 per cent at 1991-92 as compared to 6.84 per cent per annum from the exponential fit. This is suggestive of the vast untapped potential of wheat output.

PADDY OUTPUT

The computed parameters of the different growth curves for paddy output in the selected districts, namely, Ludhiana and Amritsar of the study area are presented in Table 4.2.

Based upon the criteria of highest $\overline{R}^2$ value, like wheat crop, log quadratic curve provides the best fit for the paddy crop—another cereal crop considered for the present study for both the districts under considerations, namely, Ludhiana (PHL) and Amritsar (PLA). Rather if we go by the lowest C-value criterion, Gompertz fit seems to be reasonably close to those of the log quadratic fit. Moreover, the value of $\overline{R}^2$ from the Gompertz fit was comparable to that of the log quadratic fits. There is, therefore, no reason to accept log quadratic fits as against the Gompertz fits for paddy output too in both the districts under considerations. The comparison of the reported output of paddy with that of the estimated output from the

Table 4.2: Estimated growth curves for paddy output

Model(s)	Estimated value of the regression coefficients for			
	K	a/log a	b	c
		Ludhiana		
SLM		2.315	+0.225‡ (25.472)	
LIM		0.653	–0.033‡ (7.738)	
LQM		3.425	+0.446‡ (9.736)	–0.006‡ (2.723)
MQM		3.281	+0.325‡ (29.893)	–0.006‡ (2.723)
LCM	1238.967	–2.477	0.1509	
GCM	2.936	–2.229	0.782	
		Amritsar		
SLM		5.552	+0.079‡ (12.956)	
LIM		0.216	–0.003‡ (11.647)	
LQM		4.299	+0.170‡ (6.188)	–0.004‡ (2.679)
MQM		5.644	+0.099‡ (15.131)	–0.004‡ (2.682)
LCM	1133.088	–1.014	+0.056	
GCM	3.219	–0.965	+0.948	

The computations of first four growth curves are in natural (base e) logarithmic while that of last two in base 10.
Figures in parentheses are the 't' value of the respective parameters.
‡ implies significance at 0.01 level.
† implies significance at 0.05 level.
* implies significance at 0.10 level.
ns not significant.

Gompertz fits are reported in Fig. 4.2. The value of log a is negative and that of b is less than unity for both the districts. This implies that the rate of growth is declining towards upper asymptote. The Gompertz, with parameters implying a decreasing rate of growth, indicates a decline in the rate of growth from 10.82 per cent in 1975-76 to 5.89 per cent in 1985-86 and to 4.16 per cent in 1991-92 for Amritsar district of the study area. The corresponding decline was from 71.62

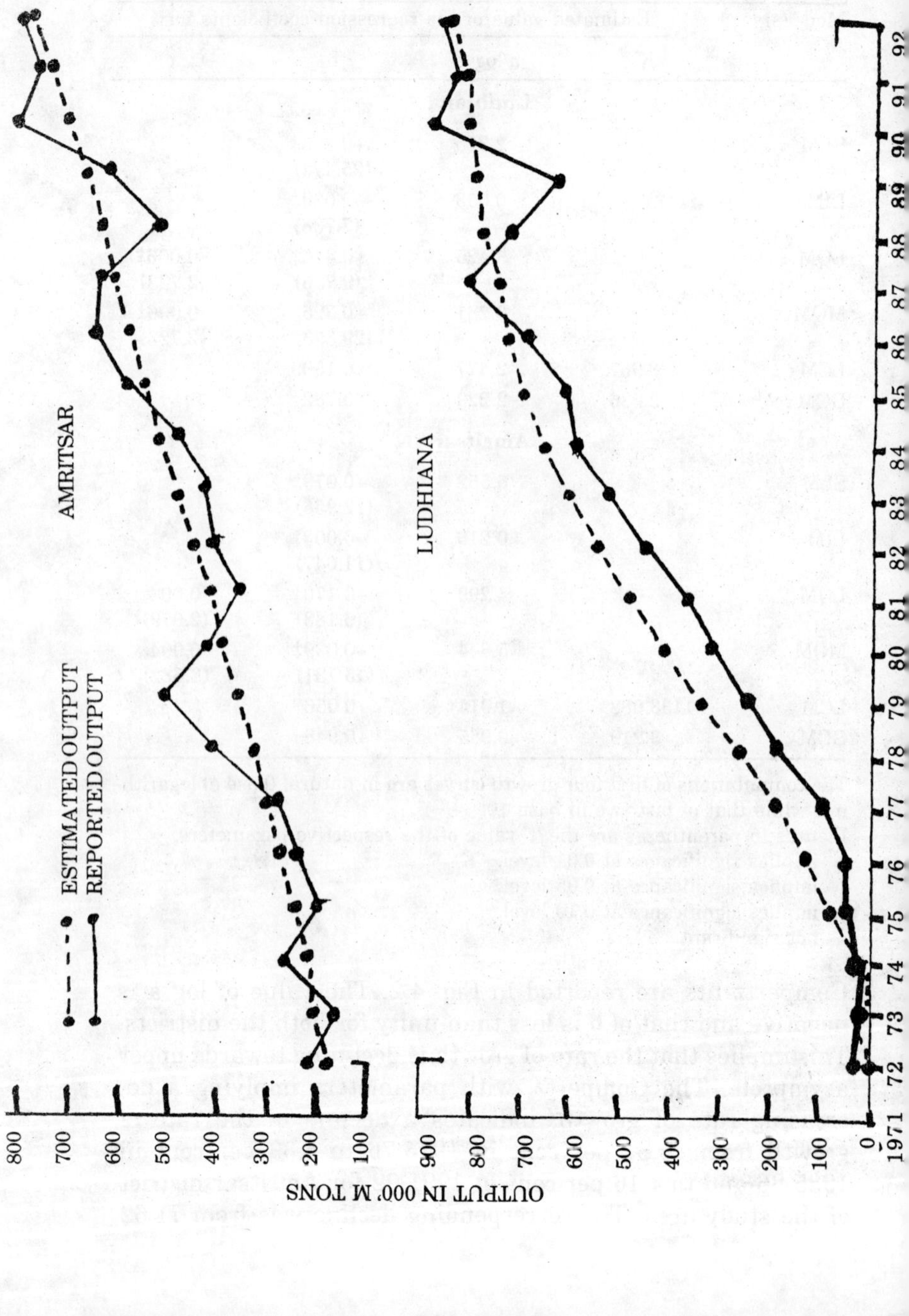
AMRITSAR
LUDHIANA
ESTIMATED OUTPUT
REPORTED OUTPUT
OUTPUT IN 000' M TONS
1971 72 73 74 75 76 77 78 79 80 81 82 83 84 85 86 87 88 89 90 91 92

per cent at 1975-76 to 4.72 per cent at 1985-86 and to 1.09 per cent at 1991-92 for the Ludhiana district.

COTTON AMERICAN

The computed parameters of the different growth curves for the output of cotton american in the selected districts, namely, Bathinda and Ferozepur, of the study area are presented in Table 4.3.

Table 4.3: Estimated growth curves for cotton American output

Model(s)	Estimated value of the coefficients for			
	K	a/log a	b	c
		Bathinda		
SLM	x	2.834	+0.086‡ (9.582)	x
LIM	x	0.344	–0.007‡ (8.224)	x
LQM	x	2.619	+0.148‡ (4.135)	–0.003 ns (1.776)
MQM	x	3.786	+0.086‡ (10.171)	–0.003 ns (1.776)
LCM	84.484	–0.516	–0.068	x
GCM	–0.643	2.121	1.015	x
		Ferozepur		
SLM	x	3.937	+0.045‡ (5.265)	x
LIM	x	0.250	–0.004 (4.785)	x
LQM	x	3.892	+0.035 ns (0.869)	–0.001 ns (0.356)
MQM	x	3.502	+0.045‡ (5.128)	–0.001 ns (0.356)
LCM	66.521	–0.271	+0.036	x
GCM	1.457	–0.006	+1.235	x

Figures in parentheses are the 't' value of the respective parameters.
The computations of first four growth curves are in natural (base e) logirithmic while that of last two are to base 10.

‡ implies significance at 0.01 level.
† implies significance at 0.05 level.
* implies significance at 0.10 level.
ns not significant.

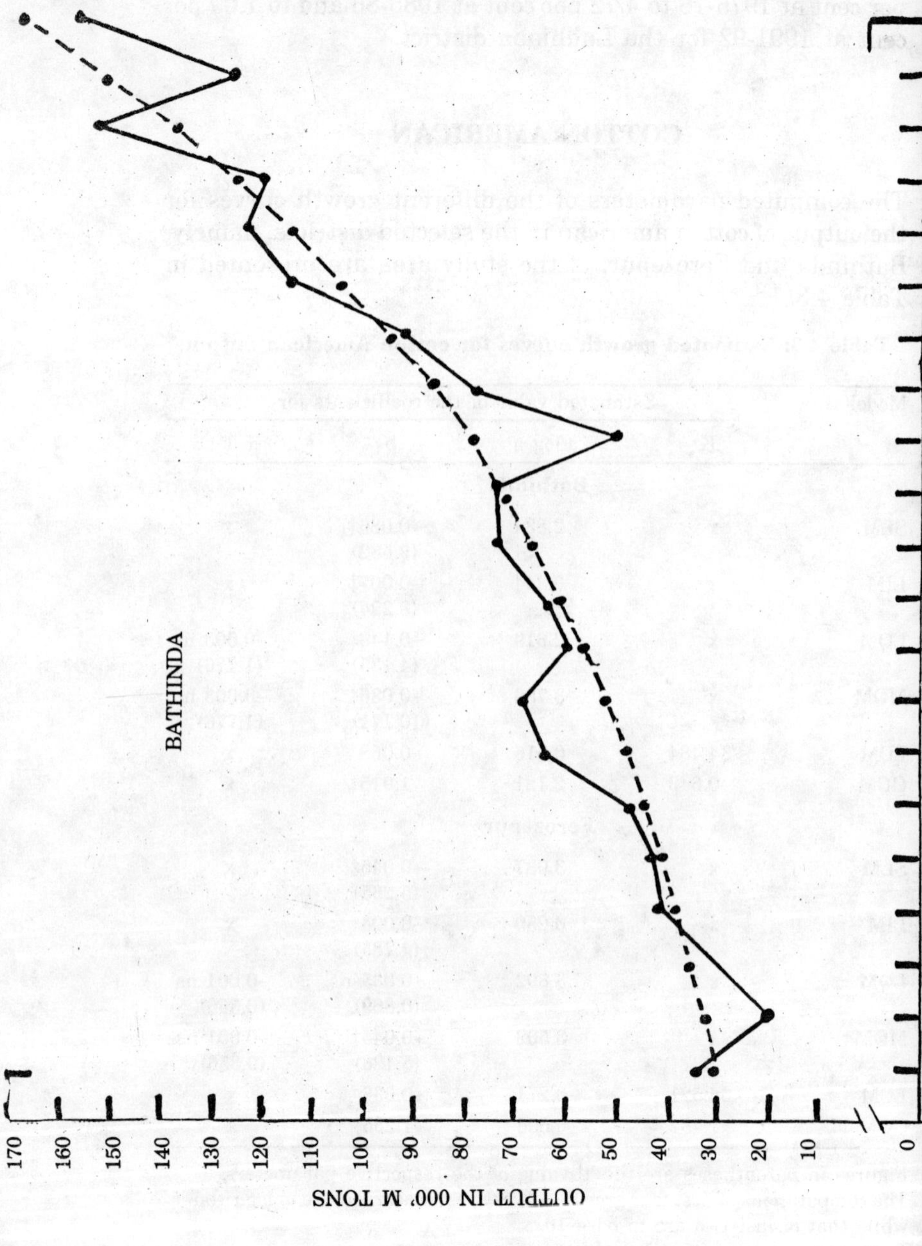
BATHINDA
OUTPUT IN 000' M TONS
170
160
150
140
130
120
110
100
90
80
70
60
50
40
30
20
10
0

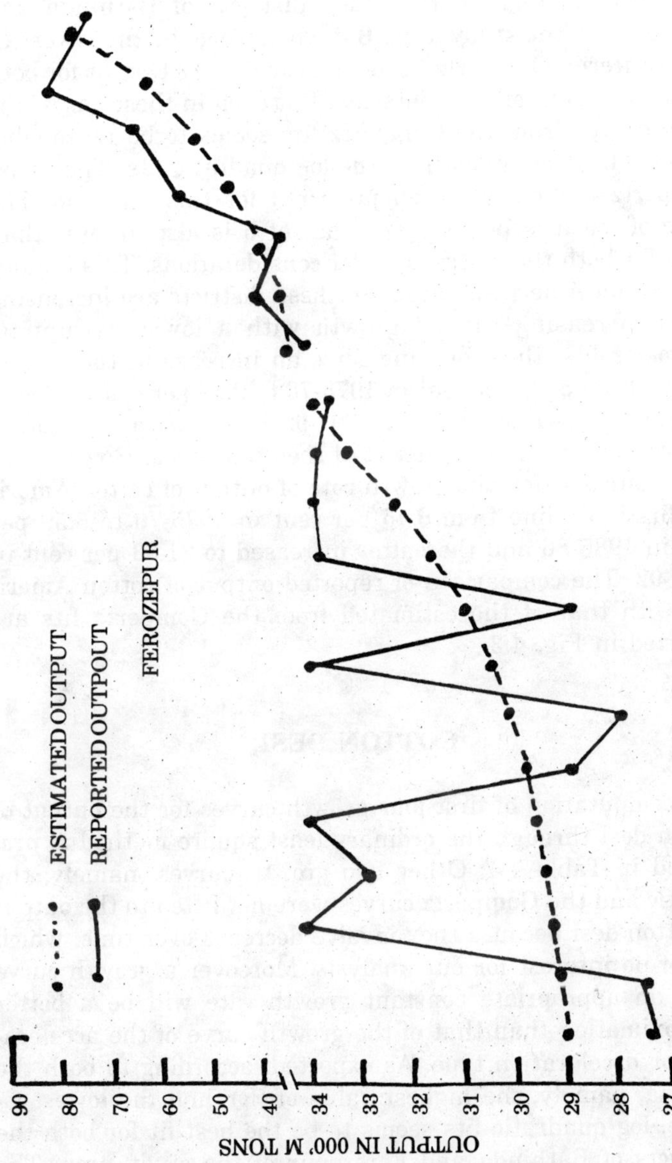

Fig. 4.3b: Comparison of estimated and reported output of cotton American (Ferozepur): 1971–72 through 1991–92

Based upon the criteria of highest $(\overline{R})^2$, like food crops of wheat and paddy, log quadratic fits gave the best fit for cotton american too in both the study districts of Bathinda and Ferozepur in the study area. But when we go by the lowest C-value criteria, Gompertz fit again provides the best fit for both the districts under consideration. But even in these cases, the value of $(\overline{R})^2$ from the Gompertz fits seems to be reasonably close to those obtained from the log quadratic fits. Therefore, Gompertz curves has been preferred for this crop too. The value of log a is positive and that of b is also greater than unity for both the districts under considerations. This implies that cotton American output in these districts are increasing at an increasing rate of growth with a lower asymptote. Gompertz fits, therefore, indicates an increase in the rate of growth from 8.09 per cent in 1975-76 to 9.45 per cent in 1985-86 and to 10.41 per cent in 1991-92 in the output of cotton American in Bathinda district of the study area. However, in Ferozepur district, the growth rate of output of cotton American firstly decline from 6.45 per cent in 1975-76 to 5.59 per cent in 1985-86 and thereafter increased to 21.33 per cent in 1991-92. The comparison of reported output of cotton American with that of the estimated from the Gompertz fits are reported in Fig. 4.3

COTTON DESI

The computation of first four growth curves for the output of cotton desi through the ordinary least square method is presented in Table 4.4. Other two growth curves, namely, the logistic and the Gompertz curves were not fitted to the output of cotton desi because they involve decrease over time, which are of no interest for our analysis. Moreover a growth curve with an appropriate constant growth rate will be a better approximation than that of the growth curve of the acceleration or deceleration type. As expected, according to both the criteria, namely, the highest value of $(\overline{R})^2$ and the lowest C-value, log quadratic fits seems to be the best fit for both the districts of Bathinda and Ferozepur of the study area. The output of cotton desi, therefore is declining with a growth rate

Table 4.4: Estimated growth curves for cotton desi output

Model(s)	Estimated value of the coefficients for			
	K	a/log a	b	c
		Bathinda		
SLM		3.446	–0.094‡ (8.007)	
LIM		0.243	–0.018‡ (6.213)	
LQM		3.105	–0.004 ns (0.087)	+0.005‡ (2.268)
MQM		2.657	–0.094‡ (8.930)	+0.005‡ (2.268)
		Ferozpur		
SLM		3.096	–0.154‡ (9.237)	
LIM		1.216	–0.102 ns (0.917)	
LQM		3.462	–0.259‡ (3.838)	+0.005 ns (1.597)
MQM		1.399	–0.154‡ (9.649)	+0.005 ns (1.597)

The computations of growth curves are in natural (base e) logarithmic.
Figures in parentheses are the 't' value of the respective parameters.
‡ implies significance at 0.01 level.
† implies significance at 0.05 level.
* implies significance at 0.10 level.
ns not significant

of 9.5 per cent in Bathinda and 15.4 per cent in Ferozepur district. The comparison of the reported output with that of the estimated output at the constant growth of decline is reported in Fig. 4.4.

SUGARCANE

The computed parameters of the different growth curves for sugarcane output in the selected districts, namely Gurdaspur, and Patiala of the study area are presented in Table 4.5. Based upon the criteria of highest $(\overline{R})^2$ value, semi logarithmic fits gave the better fit for sugarcane output in both the dis-

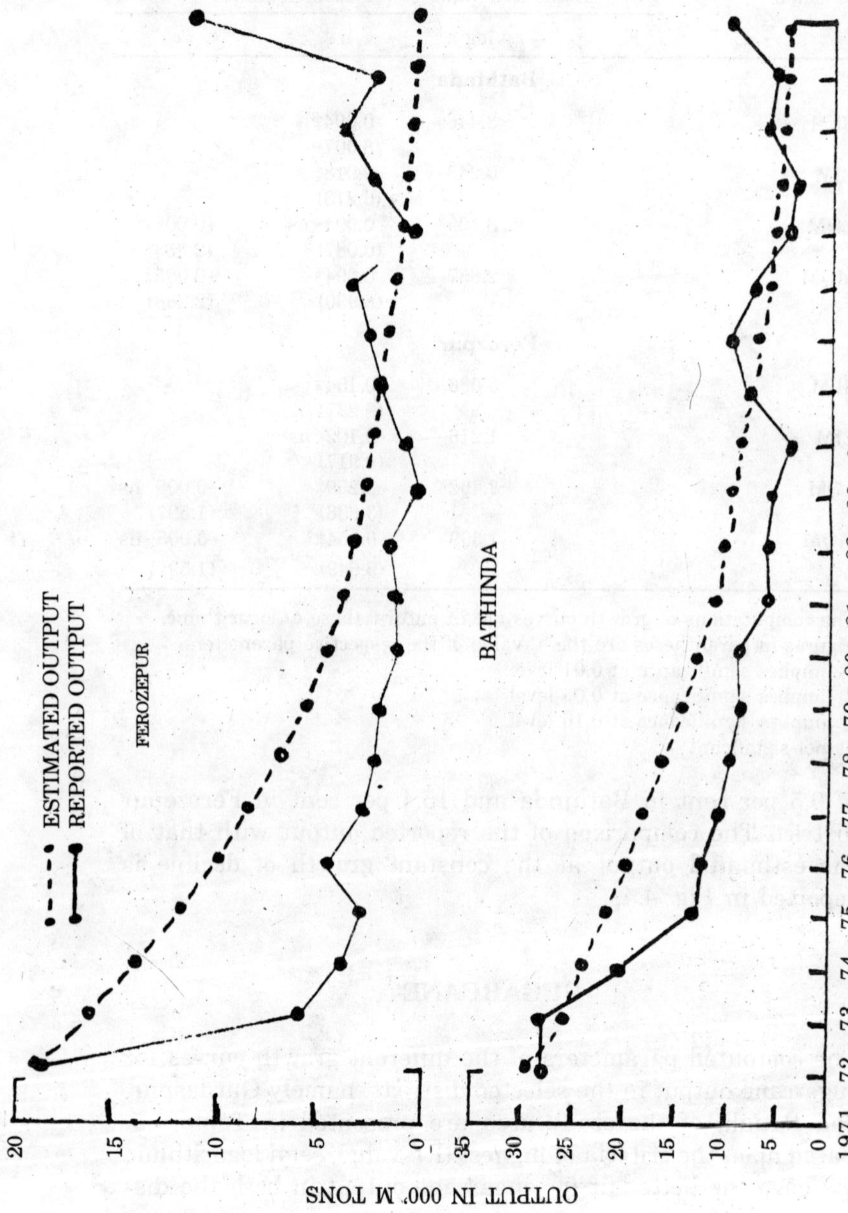

Fig. 4.4: Comparison of estimated and reported output of cotton desi: 1971–72 through 1991–92

Table 4.5: Estimated growth curves for sugarcane output

Model(s)	Estimated value of the coefficients for			
	K	a/log a	b	c
		Gurdaspur		
SLM	x	4.220	+0.024‡ (2.485)	x
LIM	x	0.237	–0.001‡ (2.417)	x
LQM	x	4.098	+0.059 ns (1.417)	–0.002 ns (0.857)
MQM	x	4.517	+0.024‡ (2.446)	–0.002‡ (2.857)
LCM	128.180	–0.069	+0.053	x
GCM	1.962	–0.002	1.262	x
		Patiala		
SLM	x	4.029	–0.001 ns (0.289)	x
LIM	x	0.249	–0.002 ns (0.297)	x
LQM	x	3.882	–0.039 ns (0.996).	–0.002 ns (1.096)
MQM	x	4.069	–0.003 ns (0.291)	–0.002 ns (1.096)

The computations of first four growth curves are in natural (base e) logarithmic while that of last two in base 10.
Figures in parentheses are the 't' value of the respective parameters.
‡ implies significance at 0.01 level.
† implies significance at 0.05 level.
* implies significance at 0.10 level.
ns not significant.

tricts under considerations. Even going by the criteria of lowest C-value, semi logarithmic fit seems to be better fit for the sugarcane output in Patiala district while Gompertz fit was better for Gurdaspur district output of sugarcane. Gompertz having negative value of log a and that of b less than unity, implies that the output of sugarcane in Gurdaspur district is increasing at the increasing rate of growth tending towards lower asymptote. The increase in the rate of growth was noticed from 1.09 per cent in 1975-76 to 2.04 per cent at 1985-86 and to 6.67 per cent at 1991-92. The comparison of the reported output sugarcane with that of the estimated output are presented in Fig. 4.5.

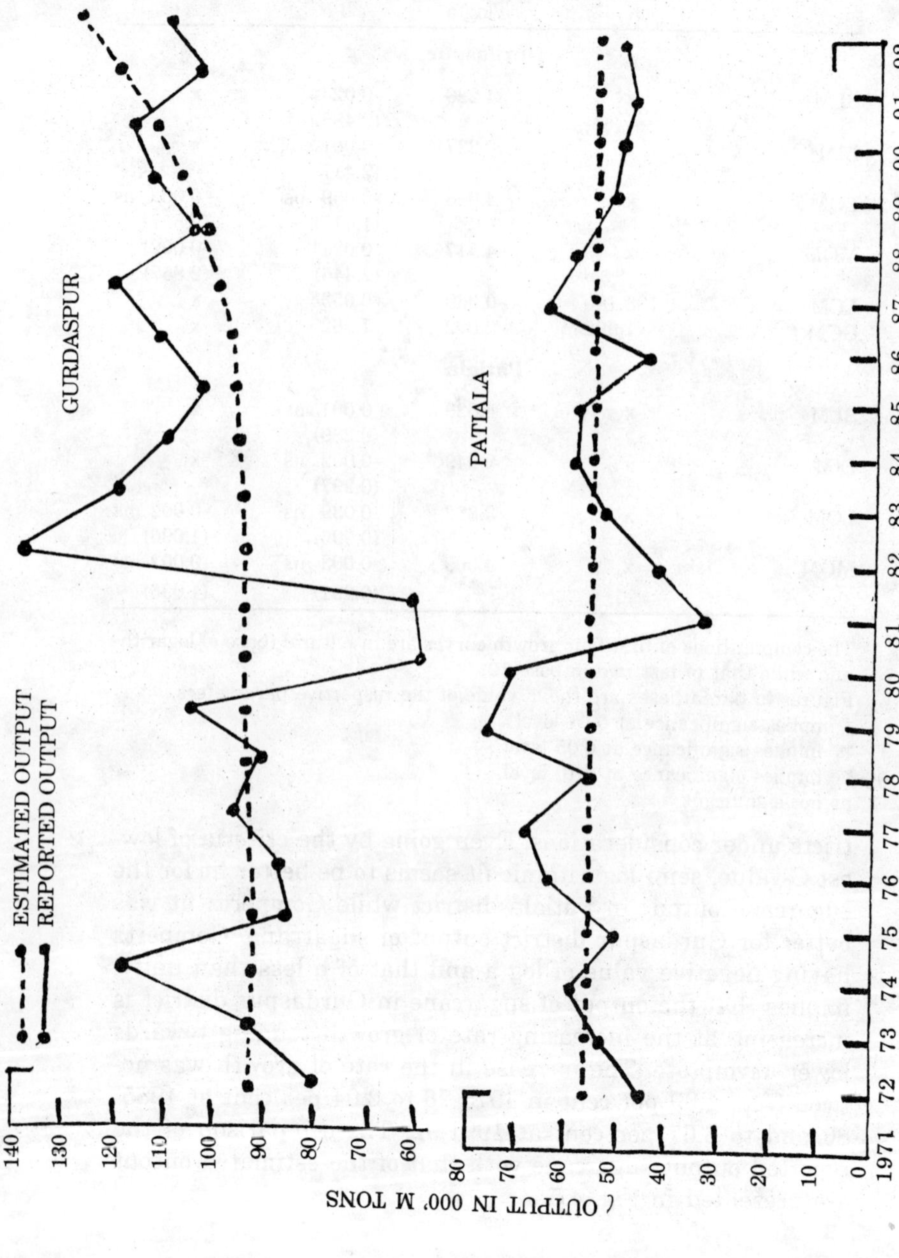

Fig. 4.5: Comparison of estimated

RAPESEED AND MUSTARD

The computation of the parameters of the different growth curves for the output of the rapeseed and mustard in Amritsar district of the study area are presented in Table 4.6.

According to the highest $(\overline{R})^2$ criteria, semi-logarithmic fit seems to be the best fit for oil seed crop of Amritsar district, namely, rapeseed and mustard. But when we go by criteria of lowest C-value, Gompertz fit always seem to be better than the log quadratic fit. Thus, we may conclude that in so far as the goodness of fit is concerned, there is no reason to accept semi logarithmic fit as against the Gompertz fit for the output of rapeseed and mustard. The comparison of the reported output with that of the estimated output from the Gompertz fit is shown in Fig. 4.6. Furthermore, the value of log a is negative and that of b is less than unity, implies that the rate of growth of output of rapeseed and mustard is declining towards upper asymptote. The Gompertz, with parameters implying declining rate of growth, indicates a decline in the rate

Table 4.6: Estimated growth curves for output of rapeseed and mustard

Model(s)	Estimated value of the coefficients for			
	K	a/log a	b	c
SLM		2.487	+0.085* (1.194)	
LIM		0.416	–0.005* (1.918)	
LQM		2.437	+0.049 ns (0.622)	–0.001 ns (0.187)
MQM		2.859	+0.035 ns (1.159)	–0.001 (0.187)
LCM	105.631	–1.028	+0.031	
GCM	1.801	–0.603	+0.973	

The computations of first four growth curves are in natural (base e) logarithmic while that of last two in base 10.

Figures in parentheses are the 't' value of the respective parameters.

‡ implies significance at 0.01 level.

† implies significance at 0.05 level.

* implies significance at 0.10 level.

ns not significant.

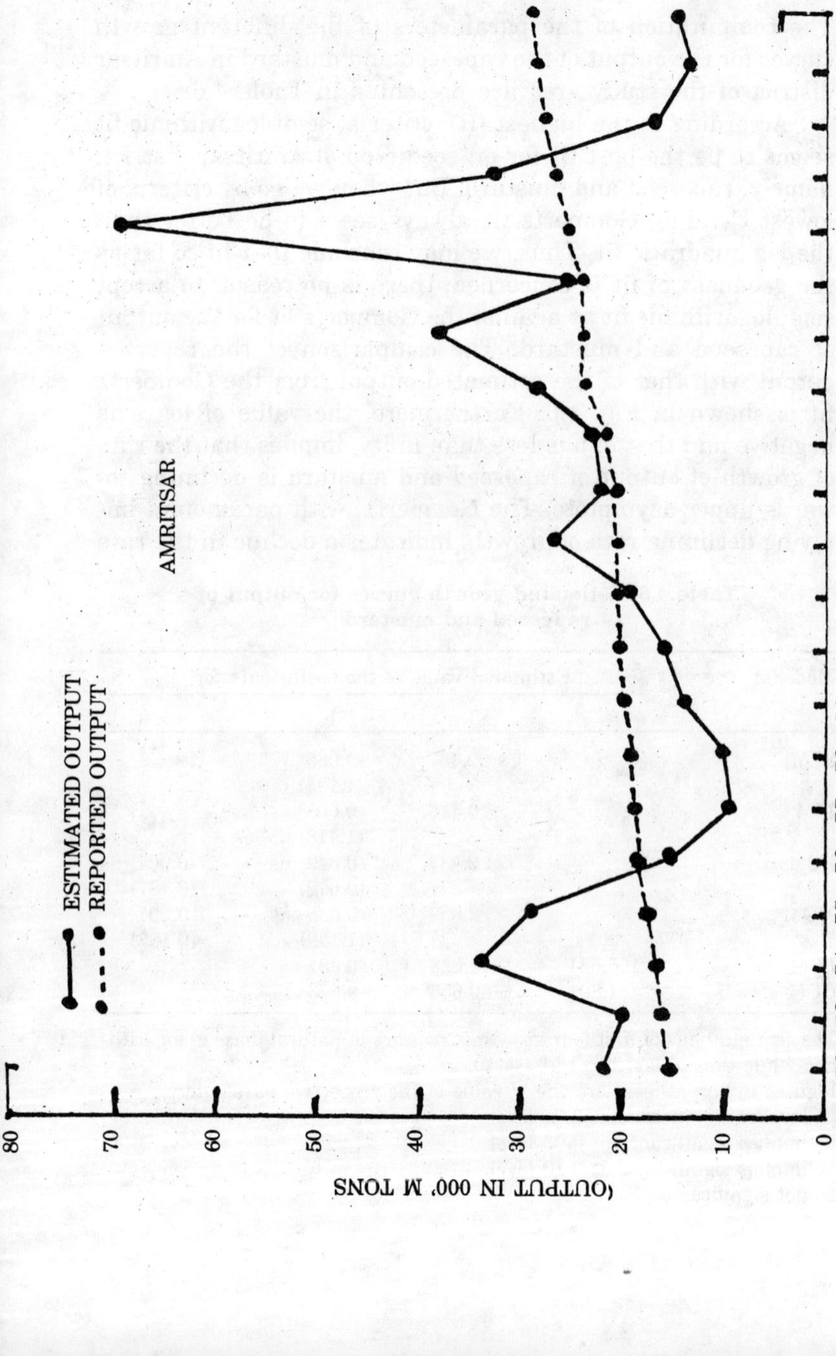

Fig. 4.6: Comparison of estimated and reported output of rapeseed and mustard: 1971–72 through 1991–92

of growth from 3.75 per cent in 1975-76 to 2.58 per cent at 1985-86 and to 2.37 per cent at 1991–92.

Our analysis, based upon the fitting of six different growth curves, namely, semi logarithmic; log inverse; log quadratic; modified log quadratic; logistic and the Gompertz curves, and then judging the goodness of fit on the basis on the highest $(\overline{R})^2$ and the lowest C-value, yield some interesting results. Firstly, the results of the fitted growth curves allow us to reject the hypothesis concerning the unsuitability of the semi-logarithmic growth curve (SLM) which is due to the existence of an upper/lower asymptotic boundary of the growth of crop production and the law of continuously declining/increasing annual rates of growth. Secondly, the growth of agriculture output reveals a smooth trend as to rule out any districts-wise wide share changes in the growth of crop production due to introduction of new agricultural technology, consisting of seed-fertilizer-water technology. Thirdly, crop-wise trend analysis suggests that it is again the wheat which shows the faster growth in output. Acceleration in the growth of crop output following the green revolution has been tapering off. In otherwords, there is yet no green revolution but still only a wheat revolution. Thus, the commonly prevailed popular belief that green revolution has succeeded only in the case of *rabi* crops in general and wheat in particular comes true atleast in case of Punjab agriculture. Indeed lower growth rate of crop output reminds us of the difficulty of sustaining a much higher growth rate of agricultural output in the foreseeable future.

REFERENCES

1. The aggregation may be of crops or over the region viz., district across states and states across nation.
2. Mention may be made of the work done by Mitra (1968); Minhas and Srinivasan (1968); Rudra (1970, 1982); Dey (1975); Vidyanathan and Srinivasan (1977); Reddy (1978); Srinivasan (1979); Dandekar (1980); Chaudhri (1981); Ray (1983); Sau (1983); Rao, Nandkarni and Deshpande (1980) and Mukherjee and Vaidyanathan (1980)
3. See the instance, V.M. Dandekar, "Introduction to seminar on Data Base and Methodology for the study of growth Rates in Agriculture," *Indian Journal of Agricultural Economics* Vol. 35, No. 2, April–June 1980.

4. In each of these functions, the parameters a,b,c .... possess specific values. Specific notations were not introduced for these on the assumption that these are self evident.
5. For instance, please see, T.N. Srinivasan, "Trends in Agriculture in India, 1949-50 to 1977-78," *Economic and Political Weekly*, Vol. 14, Nos. 30, 31 and 32, Special Number, August 1979.
6. In an ideal good fit, it is expected that the observed points should lie on both sides of the fitted curve evenly. In the extreme case we are likely to get fluctuations around the trend curve in every period and Q in that case equals (n–1), where P is zero. In such a case C= –1. On the other extreme, a high value C implies a high value of P relative to Q and little fluctuations around the trend curve. Therefore, $-1 < \beta <$ 1. For detail see, N. Bhattacharya and T. Maitra, *An Analytical of Engel Curves Based on NSS Household Budget Data from 7th to 22nd Rounds*, Technical report No. ECON/9/70, Research and Training School, Indian Statistical Institute, Calcutta, 1970.

Annex Table 4.1
Algebric sum, standard deviations and coefficient of variants of the difference between the reported and observed outputs

Crop(s)	Model-I			Model-II			Model-III		
	AS	SD	CV	AS	SD	CV	AS	SD	CV
WHL	68.21	74.59	20.78	0.00002	0.003	207.47	147.06	74.29	9.60
WLG	91.20	83.51	6.98	-0.00004	0.004	-1792.80	35.91	36.26	19.23
PHL	-450.68	130.25	-5.49	-0.04	0.14	-6.42	-37.43	58.35	29.55
PLA	26.44	16.26	43.30	0.0002	0.007	621.6	-72.39	50.56	-13.25
CDF	11.78	3.15	5.07	-0.00001	2.51	-477106.57	5.70	3.70	12.41
CDB	6.65	4.89	14.16	0.00004	0.07	4050.55	7.98	3.69	8.78
CAF	-77.33	22.92	-5.63	0.06	0.02	6.61	3.61	6.59	35.30
CAB	10.83	9.87	17.22	0.00002	0.02	1969.25	10.64	8.14	14.43
SGP	22.61	11.15	9.36	0.01	0.01	-22.10	21.09	10.65	9.62
SGG	40.09	19.14	9.47	-0.0005	0.01	-42.12	38.95	19.02	9.28
RSA	25.84	6.91	5.08	-0.00002	0.06	-5918.49	25.38	7.11	5.05

(Contd.)

Annex Table 4.1 (Contd.)

Crop(s)	Model-IV			Model-V			Model-VI		
	AS	SD	CV	AS	SD	CV	AS	SD	CV
WHL	59.47	71.28	22.80	818.81	74.71	1.73	58.04	63.13	19.58
WLG	35.72	36.26	19.24	133.13	38.51	5.50	6.15	33.42	97.80
PHL	–37.43	58.35	29.55	376.81	40.46	2.04	–233.97	60.38	–4.65
PLA	64.41	50.59	15.11	503.36	53.43	2.02	193.18	52.64	4.90
CDF	6.84	3.70	10.77	N.C	N.C.	N.C	N.C	N.C.	N.C
CDB	7.98	3.69	8.78	N.C	N.C.	N.C	N.C	N.C.	N.C
CAF	6.65	6.59	18.73	27.98	6.39	4.34	16.5	4.98	1.48
CAB	14.44	8.13	10.70	39.66	8.61	4.12	146.21	85.89	0.73
SGP	26.41	10.69	7.68	N.C	N.C.	N.C	N.C	N.C.	N.C
SGG	141.17	18.83	2.53	56.76	19.74	6.61	41.68	16.77	0.03
RSA	25.84	6.92	5.09	57.69	7.25	2.39	96.91	4.71	0.88

AS, SD and CV stands for the algebraic sum; standard deviations and coefficient of variations of the difference between the reported and observed end.

For description of crop(s) code, please refer to the text.

Annex Table 4.2
Estimated value of 'C' statistics of different trend models

Crop(s) Code	Model-I (SLM)	Model-II (LIM)	Model-III (LQM)	Model-IV (MQM)	Model-V (LCM)	Model-VI (GCM)
WHL	0.444	0.556	0.222	0.222	0.222	0.177
WLG	0.444	0.444	0.444	0.444	0.444	0.312
PHL	0.444	0.178	0.222	0.222	0.444	0.059
PLA	0.111	0.333	0.111	0.111	0.333	0.059
CDF	0.222	0.889	0.444	0.444	N.C.	N.C.
CDB	0.222	0.444	0.222	0.222	N.C.	N.C.
CAF	0.444	0.666	0.808	0.808	0.222	0.082
CAB	0.333	0.556	0.606	0.606	0.000	0.165
SGP	0.111	0.111	0.222	0.222	N.C.	N.C.
SGG	0.333	0.556	0.111	0.111	0.206	0.001
RSA	0.556	0.556	0.556	0.556	0.333	0.329

For description of crop(s) code, please see the text.

Annex Table 4.3
Estimated value of $(\overline{R})2$ statistic of different trend models

Crop(s) Code	Model-I (SLM)	Model-II (LIM)	Model-III (LQM)	Model-IV (MQM)	Model-V (LCM)	Model-VI (GCM)
WHL	0.775	0.688	0.763	0.763	0.638	0.760
WLG	0.864	0.829	0.863	0.863	0.866	0.890
PHL	0.973	0.766	0.980	0.980	0.950	0.966
PLA	0.993	0.882	0.990	0.990	0.803	0.971
CDF	0.864	0.009	0.839	0.839	NC	NC
CDB	0.978	0.676	0.822	0.822	NC	NC
CAF	0.598	0.549	0.576	0.576	0.539	0.578
CAB	0.835	0.787	0.804	0.804	0.796	0.795
SGP	0.054	0.054	0.041	0.041	NC	NC
SGG	0.223	0.212	0.211	0.211	0.151	0.219
RSA	0.229	0.120	0.076	0.076	0.027	0.217

For description of crop(s) code, please see the text.

CHAPTER V

WEATHER AND CROP PRODUCTION

Economists' interest in the study of crop-weather relations is of relatively recent origin and reflects the growing realisation that the effect of weather needs to be taken out of the observed behaviour of crop yields before the contribution of technology and inputs can be properly evaluated. However, their work on the subject shows little awareness of the research already done by agricultural scientists and agro-meteorologists, or of the painstaking effort which went into the collection of the relevant basic data.[1] Every one knows that the climatic conditions have a profound influence on crop's biological growth periods and hence yield. Yet our present knowledge is such that no one knows the exact way in which the various elements of weather control the crop's biological phases and the resulting output. This is so, not because, that the knowledge is considered unnecessary but because the problem is very intricate one and depends upon:

a. collection of a quite long time series data from a net-work of stations, the crops and the weather data which are quite precise and highly dependable and according to common plan, and
b. processing and analysis of such data in an effective and rational way.

Weather influences the various crop phases in two distinct ways. On the one hand, there is a certain range of weather elements in which the various phases can take place and, on the other hand, there is a set of optimum weather conditions which are conducive for the maximum rate of germination, tillering, elongation, thickening, ripening etc. Moreover, the temporal fluctuation in agricultural production arises from the joint effect of change in both area and yield. These changes in turn are the result of interaction between weather and non-weather factors. This chapter is an attempt to establish the relationship between crop production and weather factors. As there are number of *a priori* reasons to the same weather conditions, the present study attempts to study separately the effect of weather on changes in area and yield of crops.

ECONOMETRIC SCAFFOLDING

Weather Variables

Weather, as a determinant of crop growth, may be defined to contain a number of meteorological variables such as, rainfall, temperature, humidity, wind velocity, moisture etc.[2] Some scholars have also included weather induced variables such as crop diseases and pests in this category.[3] All these variables do affect production in agriculture but accounting for all these is a very complex task. A number of methods have been utilised in the past for evaluating the impact of weather on production. While some work has been done by agricultural scientists (notable examples being Harold Mann;[4] Kalamkar and Satakopan,[5] and Kanitkar [6]), the main centre of research on crop-weather relations during the past half-a-century has been the Indian Meteorological Department (IMD). To encourage a proper scientific study of the relation between weather and plants a separate division of Agricultural Meteorology was started in 1932 and its pioneering work has been widely recognised. During infancy, the Division was mainly concerned with macro analysis of distribution and variability of rainfall, and estimation of evaporation and crop weather requirements as well as with micro studies of climatic variables near the ground to improve the understanding of the environment affecting plant growth.[7] In 1945, the Indian Council of Agricultural Research (ICAR) launched an All India Co-ordinated Crop Weather Scheme (AICES). Under this scheme, specialised meteorological observatories were set up for systematic recording of crop weather observations on paddy, wheat and jowar, at a network of selected experimental farms through out the country. These observatories were later extended to sugarcane and cotton as well. The objective of the scheme was to 'formulate, in quantitative terms, the effect of different growth factors on the growth and yield of crops under observation. . . .'[8]

Earlier work undertaken on crop weather relation was congined to simple correlation and regression analysis. The first step towards comprehensive analysis was the application of 'multiple curvilinear regression' technique evolved by Ezekiel and Fox.[9] The starting point is the linear multiple regression of the standard type:

$$Y = a_0 + a_1X_1 + a_2X_2 + \dots\dots + a_nX_n \quad \dots \quad (1)$$

While this is a useful first approximation, the relation between the dependent variable (Y) and each of the independent variables may have different and not necessarily linear forms. The true relationship, in other words, would be of the following form:

$$Y = a_0 + a_1f_1(X_1) + a_2f_2(X_2) + \dots\dots + a_nf_n(X_n) \dots \quad (2)$$

where $f_1(X_1)$, etc., may have different forms not necessarily linear. The nature and shape of $f_1(X_1)$ are got by a process of successive approximation using freehand curves. The multiple and partial correlation coefficients are then estimated by feeding the freehand approximation of the curvilinear functions into equation (2).

In case X_1, X_2, X_3, ..., X_n denote values of the variables in different periods, n is likely to be quite large. In that situation, a large number of constants have to be evaluated from the data. This will require a long series of data for precise estimates of constants, which may not be available in practice. This method is found to explain much higher proportion of the observed variations than the linear regression. But higher values of R^2 is not a sufficient basis for selecting one model over another even when standardised and objective techniques have not been developed for curvilinear regression. In any event, unless the underlying model is firmly grounded on theoretically sound and consistent hypothesis derived from the relevant disciplines, its utility for higher understanding crop-weather relations remains doubtful, no matter how high is the value R^2 ?

Fisher developed a new statistical technique what is called the 'regression integral technique' which take into account not only the total amount of rainfall during a certain period, but also the manner in which it is distributed over the period under consideration.[10] He assumed that the effects of change in weather variable in successive period (says weeks) would not be an abrupt or erratic change but an orderly one that follows some mathematical law. He assumed that these effects are composed of the terms of polynomial function of time.

If the time period (O, T) over which the influence of the meteorological factor (r) is required to be examined, be divided into n equals parts, then linear regression equation of the yield upon the meteorological factor would be of the from:

$$\dot{Y} = C + a_1 r_1 + a_2 r_2 + \ldots + a_n r_n \quad \ldots \quad (3)$$

where r_1, r_2, - - - - r_n are the quantities of the meteorological factor in the successive intervals of time and a_1, a_2, - - - a_n are the corresponding partial regression coefficients. In the limiting case, when the length of the sub divisions is made infinitely small, the equation (3) assumes the form :

$$Y = C + \int_0^T ardt \quad \ldots \quad (4)$$

where r dt is the amount of the meteorological factor in the infinitely small interval of time dt and a in the equation (4) is a continuous function of time and represents the average benefit to the crop corresponding to an additional unit of the meteorological factor at any point of time during the period under consideration. This function is known as the regression function and the graph of it is popularly known as Fisher's Response Curve.

Now if P_1, P_2,. . . .etc. be a set of polynomial function of time orthonormal in the interval (O, T) that is,:

$$P_l \int_0^T P_m \, dt = 0 \; (l \neq m) \qquad \ldots \quad (5)$$

$$\int_0^T P_m^2 dt = 1 \qquad \ldots \quad (6)$$

where l and m are any two positive integers, the meteorological factor r can always be expressed as :

$$r = A_0 P_0 + A_1 P_1 + \text{- - -} A_{n-1} P_{n-1} \qquad \ldots \quad (7)$$

where the coefficients $A_1 A_2$, A_0 etc. are called the Meteorological Distribution Constants (MDC's). Also as there are relatively slow changes in the function a it can be satisfactorily be represented by an orthogonal polynomial of degree five.

Yield response to the MDC's (A_0, A_1 . . . etc.,) assumed to take the following polynomial form :

$$a = B_0 P_0 + B_1 P_1 + \ldots + B_5 P_5 \qquad \ldots \quad (8)$$

In order to estimate B_1 B_2 B_3 ....B_5, yield is regressed on A_1, A_2, ... A_5:

$$Y = B_0 A_0 + B_1 A_1 + \ldots + B_5 A_5 \qquad \ldots \quad (9)$$

which holds good in the period (O,T)

However the effectiveness of the regression integral techniques in relation to other remains unconfirmed.[11] Nevertheless, at the conceptual level the major limitations of this method, at any rate as applied so far, is that it estimates the effect of each meteorological factor (and its time distribution) on yield separately, but not in combination. This would require a more complicated model of the relation between yield, on the one hand, and different meteorological factors and their distributions taken in combination, on the other.

Hendricks and Scholl[12] (1943) have modified Fisher's technique. They divided the crop season into n weekly intervals and have assumed that a second degree polynomial in week number would be sufficiently flexible to express the relationship.

Mathematically, it can be expressed as:

$$A_w = a_0 + a_1W + a_2W^2 \qquad \ldots (10)$$

In particular;

$$A_1 = a_0 + 1a_1 + 1^2a_2 \qquad \ldots (11)$$

$$A_2 = a_0 + 2^1a_1 + 2^2a_2 \qquad \ldots (12)$$

and,

$$A_n = a_0 + na_1 + n^2a_2 \qquad \ldots (13)$$

Substituting the expression for A_w in equation (1), the model was obtained as:

$$Y = A_0 + a_O W^\circ X_w + A_1 W^1 X_w + a_2 W^2 X_w \qquad \ldots (14)$$

or, it can be rewritten as:

$$Y = A_0 + a_0 z_0 + a_1 z_1 + a_2 z_2 \qquad \ldots (15)$$

where Z's are generated variables and defined as:

$$Z_i = W^i X_w \ (i = 0,1,2,\ldots) \qquad \ldots (16)$$

In this model, the number of constants to be determined reduces to 4 irrespective of n.

This technique can be further modified by expressing effects of changes in weather variables on yield in a certain period as the second degree polynomial in respective correlation coefficients between yield and weather variables. This will explain the relationship in a better way as it gives appropriate weightage to different periods. We have, therefore, applied this technique by expressing the effect of changes in weather variable with period (say, a month) of the particular crop season as a linear function of the respective correlation coefficients between the yield and weather variables. Under this assumption, the model used for studying the effect becomes:

$$Y = a_0 + a_1 \frac{\sum_{i=1}^{m} r_{yw} X_w}{\sum_{i=1}^{m} r_{yw}} + a_2 \frac{\sum_{i=1}^{m} X_w^2\, r_{yw}^2}{\sum_{i=1}^{m} r_{yw}^2} \qquad (17)$$

Moreover, the trend effect was removed from the yield while calculating the correlation coefficients of yield with weather variables to be used as weights.

In the simplest form, the model becomes:

$$Y = a_0 + a_1W_i + a_2W^2_i \qquad \ldots (18)$$

TECHNOLOGICAL VARIABLES

In production analysis building, accounting for change in technology[13] is a great problem. The quantitative measurement of the effects of introduction of new technology is extremely difficult. The problem becomes more complicated when adverse weather conditions offsets the benefits of technological process. A given level of output can be obtained by combining the different resources at given level of technology. The output can be increased by increasing only some or all of the factors of production. The new technology to combine these resources may be either land saving, or labour saving or capital saving. At the aggregate level, however, it is important to attribute the variations in crop output to one or the other of such factors. An examination of the data, however, suggests that the interaction of all such factors which have contributed to changes in area and production have caused a trend like movement in these variables. For the sake of simplicity to account for changes in technology and other factors over the period of time, the time variable is used as a surrogate for these changes.

The above assumption of using time as a proxy for technology and other changes is valid only if no violent changes in technology occurred during the period under reference. This assumption stands valid in the present study as the new scientific technology consisting of seed-water-fertilizer were introduced during 1965 and its effect was felt by end of 1960s. [14] Moreover, time trend variable has been used at the disparate rate. It is linearly increasing as the dependent variable (area or yield) increases in successive years otherwise, it is assumed to be constant. The relationship between the final yield of a crop and the meteorological and technological variables affecting it is a very complex. Assuming [15] that weather and technological interactions are insignificant, we can theoretically predict the final yield of a crop in a given year as:

$$Y = a_0 + a_1 W_i + a_2 W_i^2 + a_3 T \qquad \ldots (19)$$

Where T is a linear integral valued time variable termed as technological trends.[16] The equation 19 is linear in time and weather variables. The quadratic term of weather variable 'W^2' take into account the adverse effect of weather. It introduces a single curvature in the regression surface. Yields or area changes under this relationship occur from one point to another with an optimum at $W = -(a_1/2a_2)$. Consistent results would require $a_1 > 0$ and $a_2 < 0$ (if too much or too little rain has detrimental effect, the surface should record a single maxima on the positive quadrant). In the equation (19), variation due to rainfall is superimposed over an exponential growth curve in an additive form. Since the marginal increase in area or average yield is likely to decrease with increase in W, one should expect a_2 to lie between 0 and 1.

SPECIFICATION OF WEATHER VARIABLES

Weather factors considered in the present study are rainfall; maximum, minimum and mean temperature; and mean relative humidity. The data on mean relative humidity were in percentages, therefore, it were transformed into arc-sine root proportions.

Rainfall improves the water balance and moisture conditions in the soil which in turn accelerates the absorption of nutrients by the plants and hence can be termed as an input to the growth of crop. However, if too much of rainfall occurs, this hampers the crop growth. Temperature causes evaporation of moisture from the soil and transpiration from the surface of the leaf area. Prolonged high temperature, therefore, cause loss of much needed moisture for the healthy growth of the crops. To overcome this difficulty, another climatic variable, namely, synergestic heat moisture stress[17] was constructed. It is based on the combined effect of temperature and rainfall represented by a dummy variable of the form:

$$\Psi(R/T) = \begin{cases} R^{*}T, \text{ if } T>T_{gm} \text{ and } R<R_{gm} \\ O, \text{ otherwise} \end{cases}$$

where 'gm' refer to the grand mean of the variable.

It has already been mentioned that area and yield of a crop respond differently to the same weather conditions. Therefore, we present the results of regression analysis separately relating to area and yield of different crops to the selected weather variables. As the exact nature of plant physiology is not known, it is difficult to say firmly which weather variable at which particular point of time has a favourable or a detrimental effect on the area and yield of the crop. If the knowledge of such a relationship had been known in advance, the computations would present no difficulties. In the absence of such deterministic knowledge, the only alternative left is to relate the monthly data of different climatic variables with the annual area and yield and used these correlation coefficients as the weights in the final estimates. Based on such trial, we present the results of area and yield models of different crops in the different districts selected for the study.

EFFECT OF RAINFALL

Rainfall is the most important of all the climatic factors. Delay in the timely onsets of rains and breaks in the rainy spell, that is, practically dry period, during the cropping season are the two major problems which the Indian farmers have to confront with. Even in certain areas, rains are sometimes abruptly curtailed in the monsoons season. Rain provides water to the soil which meets the irrigation requirement of the crop and is lost through evapo-transpiration from the cropped field. Reduction in the growth and yield is also related with rains. Rainy weather is always associated with the reduced evapo-transpiration from the cropped surface. At the same time, the rainy season favours plant diseases and pests. It also interferes with many farming operations such as the preparation of seedbed, sowing, harvesting and the threshing of the crop. Rain increases soil moisture status but in case of higher intensity, causes the fall in the actual soil fertility through surface run-off and leaching of nutrients beyond root zone. However, when the soil moisture deficit is there and irrigation is not possible, the deficiency or absence of rainfall causes the soil horizon to dry-up further, which hampers the supply of moisture and essential nutrients to the plants, curtails the vital growth and reproductive functions of the plant and hence reduction in the yield of the crop.

The effect of rainfall on the yield and cropped area under the selected crops in the study area are examined using Model 5.19. The results are summarised separately under two subsections, namely, the Yield Response and the Area Response.

Yield Response

The resulting estimates of the effect of rainfall on the yield of selected crops in the study area are presented in Table 5.1 through 5.6. These are discussed crop-wise as below:

Paddy

Rainfall upto a certain limit significantly and positively affects the yield of paddy in both the selected districts, namely, Ludhiana and Amritsar. Furthermore, the excess of rainfall had a detrimental, though not significant, effect on the yield of paddy crop in Ludhiana district, but has beneficial but not significant effect in Amritsar. Under this relationship, yield changes occur from one point to another with an optimum at 38.09 cm for Ludhiana district and 35.81 cm for Amritsar. The detrimental effect, though not significant, is puzzling because of the sandy loam soil condition of Ludhiana. This may, however, be due to the fact that almost cent per cent of the cropped area is timely irrigated through tubewell operated either through power or by diesel engine, though at a higher cost. Against this, net sown area irrigated in Amritsar was equally irrigated through canals and tubewells. In addition, the under ground water is salt affected in certain pockets of the district and hence not suitable for irrigation. Farmers usually irri-

Table 5.1: Estimates of weather supply functions of paddy yield

Weather Variable(s)	Intercept	Estimated value of Regression Coefficient on			$(\bar{R})^2$	Durbin-watson Statistic
		W	W^2	T		
			Ludhiana			
Rainfall	1119.917	40.839‡ (2.332)	−0.536 ns (1.088)	220.681‡ (11.342)	0.878	0.320
Maximum Temperature	−93146.100	5156.856‡ (116.054)	−69.862‡ (119.645)	165.556‡ (9.426)	0.900	0.655
Minimum Temperature	1283.488	0.678 ns (0.885)	−0.011* (1.773)	229.986‡ (12.107)	0.883	0.518
Mean Temperature	38696.200	2216.020‡ (34.791)	−32.818‡ (34.379)	221.272‡ (11.389)	0.878	0.435
Relative Humidity	4331.899	166.256‡ (15.604)	−2.150‡ (18.450)	218.202‡ (14.127)	0.923	0.636
Heat Moisture Stress	1131.153	−0.519* (1.960)	0.002 ns (1.229)	236.138‡ (13.172)	0.896	0.523

			Amritsar			
Rainfall	1403.505	1.504‡	0.021 ns	95.406‡	0.655	0.640
		(2.958)	(1.328)	(6.346)		
Maximum Temperature	64231.540	–3136.036‡	39.089‡	84.108‡	0.667	1.337
		(73.349)	(72.140)	(5.693)		
Minimum Temperature	9219.323	–916.968‡	21.862‡	90.145‡	0.637	0.920
		(18.871)	(20.206)	(5.847)		
Mean Temperature	1488.990	–6.703	0.093	88.066‡	0.604	0.810
		(0.327)	(1.165)	(5.471)		
Relative Humidity	5587.586	182.018‡	–1.892 ns	81.305‡	0.674	1.138
		(20.087)	(0.589)	(5.560)		
Heat Moisture Stress	1535.188	–2.822†	0.012 ns	82.700‡	0.613	1.087
		(2.451)	(0.514)	(5.196)		

‡ implies significance at 0.01 level
† implies significance at 0.05 level
* implies significance at 0.10 level
ns implies not significant

Table 5.2: Estimates of weather supply functions of wheat yield

Weather Variable(s)	Intercept	Estimated value of Regression Coefficient on				Durbin-watson Statistic
		W	W^2	T	$(\bar{R})^2$	
			Ludhiana			
Rainfall	2759.002	–2.277† (2.4924)	0.184 ns (0.3093)	92.403‡ (6.558)	0.696	0.818
Maximum Temperature	2185.130	7.981‡ (5.504)	–0.122‡ (4.239)	93.819‡ (6.505)	0.682	1.213
Minimum Temperature	2937.940	–69.623† (2.906)	4.078 ns (1.293)	97.406‡ (6.461)	0.653	1.366
Mean Temperature	2946.914	–2.578‡ (5.272)	0.067 ns (1.234)	91.133‡ (5.820)	0.625	1.205
Relative Humidity	2339.652	27.882† (2.620)	–0.371‡ (3.998)	102.841‡ (6.452)	0.612	1.292

			Amritsar			
Rainfall	2378.043	–490.945‡ (4.631)	59.931 ns (1.077)	102.596‡ (4.582)	0.567	2.214
Maximum Temperature	–4342.605	379.528‡ (4.502)	–5.427‡ (3.368)	101.784‡ (3.828)	0.389	1.954
Minimum Temperature	1603.803	–106.936† (2.148)	4.677 ns (0.452)	104.133‡ (4.472)	0.531	1.841
Mean Temperature	5915.561	–513.862‡ (3.390)	15.508 ns (1.125)	103.865‡ (3.812)	0.858	2.094
Relative Humidity	–6255.188	305.883‡ (8.919)	–2.811‡ (10.497)	118.362‡ (4.529)	0.610	2.200

‡ implies significance at 0.01 level
† implies significance at 0.05 level
* implies significance at 0.10 level
ns implies not significant

Table 5.3: Estimates of weather supply functions of cotton American yield

Weather Variable(s)	Intercept	Estimated value of Regression Coefficient on W	W^2	T	$(\bar{R})^2$	Durbin-watson Statistic
			Ferozpur			
Rainfall	412.029	–2.140 ns (1.262)	0.126 ns (1.695)	–5.741 ns (0.944)	0.106	1.446
Maximum Temperature	235.787	–7.777 ns (0.961)	0.130 ns (0.761)	5.658 ns (1.019)	0.077	1.651
Minimum Temperature	–232.748	69.512 ns (0.595)	–1.894 ns (0.379)	–3.336 ns (–0.539)	0.148	1.273
Mean Temperature	492.850	–8.903 ns (1.625)	0.229 ns (1.596)	–4.616 ns (0.744)	0.151	1.381
Relative Humidity	–1738.366	77.012‡ (18.747)	–0.681‡ (19.544)	–6.297 ns (1.044)	0.689	1.425

			Bathinda			
Rainfall	354.175	16.620‡ (7.048)	−1.084† (2.779)	0.625 ns (0.276)	0.636	1.396
Maximum Temperature	356.179	4.233 ns (0.891)	−0.064 ns (0.062)	−3.837 ns (0.985)	0.075	1.526
Minimum Temperature	−3910.866	371.448 ns (1.029)	−7.967 ns (0.544)	−0.729 ns (0.238)	0.336	1.245
Mean Temperature	499.440	−8.588 ns (1.040)	0.167 ns (1.143)	−2.863 ns (0.709)	0.155	1.491
Relative Humidity	223.916	13.857‡ (15.134)	−0.135‡ (19.474)	1.466 ns (0.563)	0.519	1.767

‡ implies significance at 0.01 level
† implies significance at 0.05 level
* implies significance at 0.10 level
ns implies not significant

Table 5.4: Estimates of weather supply functions of cotton desi yield

Weather Variable(s)	Intercept	Estimated value of Regression Coefficient on W	W²	T	$(\bar{R})^2$	Durbin-watson Statistic
		Ferozepur				
Rainfall	363.469	0.166‡ (4.699)	−0.354 ns (0.668)	−7.330 ns (1.173)	0.339	1.053
Maximum Temperature	29.876	5.926 ns (0.349)	−0.117 ns (0.963)	−5.585 ns (0.938)	0.657	1.189
Minimum Temperature	5274.299	488.038 ns (1.194)	−12.109 ns (1.257)	−12.797† (2.580)	0.344	1.207
Mean Temperature	150.603	6.175‡ (7.296)	−0.132 ns (0.709)	−7.865‡ (6.125)	0.656	1.209
Relative Humidity	−2913.408	131.695‡ (31.215)	−1.324‡ (34.824)	1.137 ns (0.229)	0.645	1.891

Bathinda

Rainfall	323.801	17.785‡ (3.783)	–1.491‡ (3.619)	–8.115 ns (1.592)	0.421	1.612
Maximum Temperature	137.718	27.976 ns (1.051)	–0.391 ns (0.355)	–6.330 (1.197)	0.275	1.582
Minimum Temperature	474.746	10.419 ns (0.811)	–0.228 ns (0.563)	–7.491 ns (–1.442)	0.237	1.309
Mean Temperature	–804.356	87.541* (1.969)	–1.659 ns (0.396)	–3.404 (0.685)	0.351	1.879
Relative Humidity	1298.171	–38.992‡ (12.924)	0.406‡ (14.777)	–11.547† (2.350)	0.671	1.532

‡ implies significance at 0.01 level
† implies significance at 0.05 level
* implies significance at 0.10 level
ns implies not significant

Table 5.5: Estimates of weather supply function of sugarcane yield in Patiala

Weather Variable(s)	Intercept	Estimated value of Regression Coefficient on			$(\bar{R})^2$	Durbin-watson Statistic
		W	W^2	T		
Rainfall	3275.742	107.116‡ (3.372)	–2.835* (1.982)	195.061‡ (4.735)	0.623	1.020
Maximum Temperature	10593.190	308.469‡ (4.410)	–3.169 ns (0.058)	221.989‡ (5.249)	0.602	1.527
Minimum Temperature	4081.885	38.418† (2.457)	–0.648 ns (1.807)	235.004‡ (5.450)	0.587	1.400
Mean Temperature	10097.930	508.525‡ (7.721)	–8.173 ns (0.994)	215.951‡ (4.941)	0.575	1.405
Relative Humidity	5480.038	84.160‡ (5.329)	–0.812‡ (7.790)	204.914‡ (5.358)	0.675	1.110

‡ implies significance at 0.01 level
† implies significance at 0.05 level
* implies significance at 0.10 level
ns implies not significant

Table 5.6: Estimates of weather supply functions of yield of rapeseed in Amritsar

Weather Variable(s)	Intercept	Estimated value of Regression Coefficient on			$(\bar{R})^2$	Durbin-watson Statistic
		W	W^2	T		
Rainfall	513.409	5.012‡ (3.049)	0.106‡ (2.622)	48.128‡ (4.533)	0.587	1.324
Maximum Temperature	57766.980	–3093.254‡ (–81.302)	41.732 ns (1.229)	53.558‡ (4.620)	0.508	1.331
Minimum Temperature	1479.747	–75.057 ns (0.579)	1.518 ns (1.062)	55.540‡ (4.828)	0.515	0.974
Mean Temperature	1986.226	–110.059‡ (17.029)	1.829 ns (1.370)	55.099‡ (5.669)	0.654	1.397
Relative Humidity	–2237.314	94.452‡ (10.750)	–0.811‡ (12.325)	61.546‡ (5.679)	0.569	1.553

‡ implies significance at 0.01 level
† implies significance at 0.05 level
* implies significance at 0.10 level
ns implies not significant

gated their paddy crops with the mixed water (canal + tubewell) as their requirement was not met fully with the canal irrigation only. Hence the excess of rainfall had beneficial, though not significant, effect on the yield of paddy in Amritsar.

The beneficial effect of rainfall was noticed throughout the growing crop season except during the panicle initiation and fertilization stage. The pronounced beneficial effect was, however, noticed in the month of August, that is, during tillering stage, in both the districts, and to some extent in the month of October in Amritsar district only. This was mainly due to the cultivation of Basmati paddy crop on the larger area which was sown late in the month of July-August in Amritsar district.

All this is suggestive of the fact that as for paddy crop is concerned, the available irrigation (both the canals and tubewells) is not sufficient enough to overcome the absence of rainfall and the available irrigation supplements rainfall and is not its complete substitute. This is perhaps due to high water requirement of the crop, which varied from 37 to 75 acre-inches of water depending upon soil and weather conditions. No doubt, cent per cent in Ludhiana and 49 per cent in Amritsar of the net sown area irrigated is by tubewells, but inadequate and erratic supply of power has added to the woes of the paddy growers. The power supply for tubewells is hardly available for five to six hours a day and that too is marred by constant tripping, breakdowns etc. The low voltage of power added to this problem as the electric motors do not function if the voltage dips. In addition to this, ground water is receding down steadily, about 20 cm per year. Apparently, the discharge per unit of time from the tubewells get affected unless corrective measures are taken. The general belief, therefore, that the technological innovations, such as, assured irrigation can ensure the freedom from the clutches of the vagaries of rainfall, is a myth as far as paddy crop is concerned in Punjab.

Wheat

Contrary to the results for the paddy crop, wheat grain yield experienced a detrimental effect on the growing season rainfall in both the selected districts, namely, Ludhiana and Amritsar under considerations. The reason is quite obvious. The low to moderate rainfalls due to North-East monsoons during the growing season does not leach hence affect the tillering of the plant and also the nutrient uptake ability of the plants, which ultimately lowers the yield level. This detrimental effect had been much more pronounced in the month of December. The reason is quite obvious. Wheat is generally sown by the end of October or early November. First irrigation at the crown root initiation stage needs to be applied by the end of November. And any rains after the first irrigation is bound to affect the crop in the saline areas of Punjab because wheat is sown fully under irrigated conditions. Furthermore, the excess of rainfall had not any significant effect on the yield of wheat crop. Under this relationship, yield of wheat changes occur from one point to another with an optimum at 6.19 cm in Ludhiana and 4.09 cm in Amritsar. During the growing season of wheat, the volume of rains is almost very low and the deficiency of water arising out of the moderate rainfall is fully compensated by the assured irrigation, and, therefore, probably the effect of excess rain does not show up.

Cotton American

The resulting estimates of yield response presented in Table 5.3 reveals that the time trend variable—a proxy for technology, turned out to be non significant and carried a negative sign, for both the districts, namely, Ferozepur and Bathinda under considerations. In Ferozepur district, the coefficient of rainfalls carried a negative sign but was non significant. The excess of rainfall again had no significant effect. Under this relationship, yield changes from one point to another with an optimum at 8.49 cm. This detrimental effect was observed for most of the growing season rains with the exception of July; June and November rains. In these months, rainfall had a beneficial effect on the yield of cotton American. Highest detrimental effects were noticed due to October rains followed by September rains. On the otherhand, in Bathinda district, rainfall had a beneficial effect on the yield of cotton american. However, the excess of rainfall had a detrimental effect because of the reason discussed below. Under this relationship, yield level changes from one point to another with an optimum at 7.67 cm. Pronounced beneficial effect was noticed for the month of July followed by June. This period coincides with the early growth and development of the crop. The most detrimental effect was noticed for the month of October followed by September, which was the critical period for fruiting casused flooding of the crop, resulting in inudation of the crop in low lying areas. Moreover, this rains during peak fruiting period were not congenial for boll development and high respiration and transpiration resulted in increased water loss from the plant. This lead to slow drying and wilting of cotton plant, high shedding of the reproductive parts and heavy attack of insect-pests-diseases on the crop such as incidence of bacterial leaf; alternate leaf spots; bollworms; whitefly; Jassids on the crop and hence reduced yield.

Cotton Desi

Like cotton American, time trend variable—a proxy for technology, turned out to be negative but non significant for both the districts, namely, Bathinda and Ferozepur, under consideration. Further rainfall had a beneficial effect on the yield of cotton desi. This satisfies the usual requirement of moisture during germination to reproduction stage, that is, flowering because crop is generally sown under rainfed conditions. However, the excess of rainfall had a detrimental effect on the yield of cotton desi, although it was not significant for Ferozepur. The excess of rains develops high humidity which encourages the attack of pests, insects and diseases on the crop and hence reduced yield level. Under this relationship, yield changes from one point to another with an optimum at 6.73 cm and 5.96 cm for Ferozepur and Bathinda respectively. Pronounced beneficial effect had been noticed for April rains followed by July rains in Bathinda district and the effect was most detrimental for the month of October rains because of the reason already discussed. Likewise, in Ferozepur district, the pronounced beneficial effect was noticed for April rains and most detrimental for September rains followed by October rains.

Sugarcane

Time trend variable turned out to be a positively significant factor affecting the yield of sugarcane in the study area. Furthermore, rainfall had a beneficial effect which satisfies the high requirement of moisture during germination to reproduction stage. However, the excess of rains had a detrimental effect on the yield of sugarcane because high rains develops high humidity which favours the development of various diseases, pests and insects which attack the crop and hence reduced yield level. Under this relationship, yield level changes from one point to another with an optimum at 18.89 cm in the study area. The beneficial effect of rainfall was noticed almost throughout the growing season except during the month of February. This is because the irrigated sugarcane crop is mostly sown after a preliminary heavy pre sowing irrigations (*rauni*) and rains immediately after sowing effect the germination of the crop and hence growth and yield. On the otherhand, the pronounced beneficial effect of rains was noticed during the month of June followed by July, March and April. Infact beneficial effect was noticed throughout the growing season. Summer being drier and hotter and the soil slightly coarse, the crop needs water more frequently. The maintenance of optimum soil moisture during all the stages of growth is one of the essential requisites for obtaining higher yield. Total water requirements of sugarcane crop for optimum growth varies from 200 to 300 cm depending upon the type of soil, seasonal conditions, the variety grown, the method of planting and the level of manuring and fertilization etc. Thus, the general belief that technological innovations can ensure the freedom from the clutches of the vagaries of rainfall is a myth again as for sugarcane crop is concerned.

Rapeseed and Mustard

Rainfall significantly and positively effect the yield of rapeseed and mustard crop in the study area. Furthermore, the excess of rainfall too have significantly beneficial effect on the crop yield. The reason is quite obvious. Because of low water requirement, the crop is mainly raised under un-irrigated conditions and that too on the marginal land. This beneficial effect of light showers had been noticed throughout the growing season of the crop, but had been more pronounced in the month of December, that is, at the flowering and or pod formation stage. Under this relationship, the crop yield changes from one point to another with an optimum at 23.64 cm in the study area.

AREA RESPONSE

The resulting estimates of the effect of rainfall on the area under selected crops in the study area are presented in Table 5.7 through 5.12. Like yield response, these are discussed crop-wise as below:

Paddy

Pre sowing months (of May, June and July) rainfall had a significantly beneficial effect on the area under paddy crop in the study area. However, the effect of excess of rainfall was otherway around in the study area. This is simply due to the strong irrigation base of the study area where almost cent per cent of the cropped area was irrigated, either by tubewells or by Government canals. Under this relationship, paddy area changes from one point to another with an optimum at 6.48 cm for Ludhiana district and 7.42 cm for Amritsar.

Pronounced beneficial effect was noticed for the May rains followed by July rains in both the districts, that is, the times of field preparations and transplanting. These results obtained are inconsistent with the practical situation prevailing in the study area. The crop is raised (wet system) on a land which is ploughed thoroughly and puddled with 3 to 5 cm of standing water in the field. The optimum depth of puddling is found to be around 10 cm in the clay and clay-loam types of soils. The primary objective is to obtain a fine seedbed for the seedlings to establish themselves faster and to reduce the downward percolation of water and thus to minimize the leaching losses of nutrients and thereby increased availability of plant nutrients by achieving reduced soil conditions which facilitate better availability of nutrient elements, to incorporate the weeds and stubble into the soil and to minimize the weed problems.

Wheat

Contrary to paddy crop, the area under wheat crop showed a detrimental effect of sowing season rainfall in the study area (significant for Ludhiana and non-significant for Amritsar district) though the excess of rain had non-significant effect. Under this relationship, the wheat area changes from one point to another with an optimum at 3.94 cm for Ludhiana and 2.64 cm for Amritsar district of the study area. However, the pronounced beneficial effect of September rains in Amritsar and of November rains in Ludhiana district was noticed. The reason is quite obvious. Paddy in Amritsar district is harvested and marketed in late August and or early September. And any rains during this period make the field ready for wheat sowing as it leaves much moisture in the soil. Moreover, Amritsari wheat growers prefered early sowing of wheat because of the experience of higher returns from early sown wheat.

Cotton American

Area under cotton American declined significantly in Ferozepur while in Bathinda district, it showed a positive upward trend. Furthermore, in both the districts of the study area, rainfall had a beneficial effect, though the excess of rains had a detrimental effect. Under this relationship, area under cotton american changes from one point to another with an optimum at 2.59 cm in Bhatinda and 3.04 cm in Ferozepur. Pronounced beneficial effects have been noticed for March rains in Ferozepur and of April rains for Bathinda district, being the pre sowing months and the crop is mainly raised after a preliminary heavy irrigation (*rauni*). These results are in conformity with the prevailing practices adopted in the study area.

Table 5.7: Estimates of weather function of area under paddy

Weather Variable(s)	Intercept	Estimated value of Regression Coefficient on W	W²	T	$(\bar{R})^2$	Durbin-watson Statistic
			Ludhiana			
Rainfall	–45.980	5.069‡ (3.071)	–0.391‡ (3.148)	13.448‡ (15.882)	0.933	0.374
Maximum Temperature	1985.586	95.030‡ (38.615)	–1.115 ns (1.001)	14.075‡ (17.543)	0.939	0.366
Minimum Temperature	–345.856	37.972‡ (21.792)	–0.813 ns (0.430)	13.055‡ (16.510)	0.941	0.482
Mean Temperature	–137.937	4.201‡ (13.098)	–0.068 ns (1.390)	14.014‡ (17.530)	0.940	0.532
Relative Humidity	18.986	–0.098‡ (738.842)	–0.001‡ (740.011)	11.643‡ (15.825)	0.949	0.866
Heat Moisture Stress	–29.769	–0.005* (2.093)	0.0001 ns (1.008)	13.327 (15.673)	0.932	0.369

Amritsar						
Rainfall	364.0441	+0.8993‡ (25.9370)	−0.0606‡ (24.0381)	12.1835‡ (12.0724)	0.8549	1.8903
Maximum Temperature	84.8586	−0.4396† (2.6933)	0.0052 ns (1.6669)	10.4110‡ (8.4315)	0.7827	1.1702
Minimum Temperature	32.4002	2.9204 ns (1.1665)	−0.0541 ns (0.6237)	10.2543‡ (8.3679)	0.7860	1.2175
Mean Temperature	−150.1568	14.8725‡ (5.1548)	−0.2457 ns (1.6851)	10.1441‡ (8.3006)	0.7872	0.9159
Relative Humidity	−94.6563	5.8942‡ (5.8766)	−0.0468‡ (4.3780)	9.8455‡ (8.4978)	0.8087	0.7825
Heat Moisture Stress	86.5351	−0.2770† (2.3862)	0.0006 ns (0.7366)	9.6176‡ (8.3406)	0.8105	0.9517

‡ implies significance at 0.01 level
† implies significance at 0.05 level
* implies significance at 0.10 level
ns implies not significant

Table 5.8: Estimates of weathers function of area under wheat

Weather Variable(s)	Intercept	Estimated value of Regression Coefficient on W	W²	T	$(\bar{R})^2$	Durbin-watson Statistic
For Ludhiana						
Rainfall	227.033	−12.6152‡ (7.2425)	1.5996 ns (1.1863)	4.2734‡ (7.3229)	0.6891	0.7419
Maximum Temperature	506.6197	−18.3917‡ (11.2361)	0.2835 ns (1.2466)	3.9888‡ (6.4158)	0.6471	0.6981
Minimum Temperature	217.8044	−0.2582 ns (0.8165)	0.0061 ns (0.2889)	3.5957‡ (5.7628)	0.6446	0.6080
Mean Temperature	263.3618	−3.1177† (2.6259)	0.0664 ns (0.8896)	3.8181‡ (6.0823)	0.6402	0.5579
Relative Humidity	223.8549	0.2232‡ (82.8720)	−0.0018‡ (83.3222)	3.8416‡ (6.1750)	0.6467	0.6718

			Amritsar			
Rainfall	191.147	−0.116 ns (.0121)	−0.022 ns (0.234)	9.834‡ (14.693)	0.922	0.747
Maximum Temperature	−2703.055	−170.427‡ (156.355)	2.776 ns (0.982)	10.471‡ (31.658)	0.981	1.874
Minimum Temperature	230.482	−2.856‡ (−6.546)	0.072 ns (1.159)	9.968‡ (15.296)	0.926	0.618
Mean Temperature	1142.249	−87.311‡ (−31.224)	1.989 ns (1.182)	10.084‡ (16.229)	0.933	1.149
Relative Humidity	−1322.808	43.590‡ (83.990)	−0.310 (80.879)	8.451‡ (16.615)	0.955	1.841

‡ implies significance at 0.01 level
† implies significance at 0.05 level
* implies significance at 0.10 level
ns implies not significant

Table 5.9: Estimates of weather supply functions of area under cotton American

Weather Variable(s)	Intercept	Estimated value of Regression Coefficient on			$(\bar{R})^2$	Durbin-watson Statistic
		W	W^2	T		
			Ferozepur			
Rainfall	118.1421	5.8447† (2.3799)	−0.9626* (2.0751)	−5.4304‡ (3.9686)	0.4299	1.8552
Maximum Temperature	97.9737	−2.7993‡ (5.7031)	0.0563‡ (6.7407)	−6.2901‡ (−4.7976)	0.4766	1.6584
Minimum Temperature	343.0690	−35.8044‡ (22.0758)	1.4526‡ (21.5806)	−6.2454‡ (4.8619)	0.4975	1.9323
Mean Temperature	151.4952	−2.5467‡ (−2.9167)	0.0581‡ (3.0751)	−5.7421‡ (−4.1522)	0.4177	1.9472
Relative Humidity	145.3533	−0.7815‡ (3.8218)	0.0052‡ (3.7511)	−5.3342‡ (3.8726)	0.4222	2.0816

Bathinda						
Rainfall	8.7035	13.4443‡ (4.8606)	−2.5962* (2.0983)	14.0902‡ (17.6738)	0.9361	1.7083
Maximum Temperature	30.7099	−5.6413‡ (19.2098)	0.1159‡ (21.3778)	12.3955‡ (18.5155)	0.9549	1.2733
Minimum Temperature	−95.1962	−15.7034‡ (13.7970)	0.5045‡ (16.6083)	14.2966‡ (19.7575)	0.9474	1.4858
Mean Temperature	12.8820	−3.9861‡ (3.6858)	0.1111‡ (5.7048)	13.6685‡ (17.6368)	0.9396	1.4447
Relative Humidity	312.4872	13.7679‡ (22.8123)	−0.1863‡ (28.3413)	11.5132‡ (20.3025)	0.9677	2.1135

‡ implies significance at 0.01 level
† implies significance at 0.05 level
* implies significance at 0.10 level

Table 5.10: Estimates of weaher supply functions of area under cotton American

Weather Variable(s)	Intercept	Estimated value of Regression Coefficient on			$(\bar{R})^2$	Durbin-watson Statistic
		W	W^2	T		
Ferozepur						
Rainfall	46.6034	–0.1683 ns (0.2524)	0.0020† (2.3221)	–0.3443‡ (4.6885)	0.5073	0.8660
Maximum Temperature	604.5223	29.6719‡ (20.2233)	–0.3931‡ (20.7279)	–6.2861‡ (4.8276)	0.5245	1.0762
Minimum Temperature	14.3985	6.1135‡ (6.3641)	–0.2666‡ (7.5066)	4.9518‡ (4.9518)	0.5664	0.0869
Mean Temperature	129.0104	5.0085‡ (6.8920)	–0.0759‡ (5.1953)	–7.6126‡ (5.9990)	0.5667	1.2994
Relative Humidity	54.2738	3.1477‡ (6.5048)	–0.0216‡ (7.9439)	–5.9842‡ (5.0665)	0.2646	0.9660

			Bathinda			
Rainfall	75.1152	−3.5909 ns (1.2712)	0.2434 ns (0.3694)	−6.8077† (4.7225)	0.4976	0.9404
Maximum Temperature	64.9121	2.4651‡ (9.7152)	−0.0565‡ (11.6317)	−6.0867‡ (4.9011)	0.6271	1.3920
Minimum Temperature	−242.9671	60.8820‡ (29.3420)	−2.8036‡ (27.0416)	−8.7533‡ (7.1773)	0.6404	1.0832
Mean Temperature	121.3329	9.1364‡ (5.2429)	−0.2773‡ (7.6813)	−6.3912‡ (5.0782)	0.6171	1.0734
Relative Humidity	382.5287	−16.4025‡ (19.0161)	0.1179‡ (17.2229)	−5.5567‡ (4.1282)	0.5620	1.2974

‡ implies significance at 0.01 level
† implies significance at 0.05 level
* implies significance at 0.10 level
ns implies not significant

Table 5.11: Estimates of weather supply functions of area under sugarcane in Patiala

Weather Variable(s)	Intercept	Estimated value of Regression Coefficient on W	W²	T	$\bar{R}^2$	Durbin-watson Statistic
Rainfall	15.758	–0.133 ns (0.261)	0.013 (1.139)	–1.078‡ (4.268)	0.505	1.216
Maximum Temperature	2.456	2.499‡ (6.387)	–0.038 ns (0.774)	–1.305‡ (5.156)	0.503	1.342
Minimum Temperature	14.960	0.083† (2.739)	–0.002 ns (1.102)	–1.238‡ (4.765)	0.476	1.052
Mean Temperature	30.740	2.723‡ (6.189)	–0.059 ns (1.385)	–0.989‡ (3.848)	0.487	1.145
Relative Humidity	10.951	–0.162† (2.334)	0.001† (2.413)	–1.115‡ (4.185)	0.448	1.120

‡ implies significance at 0.01 level
† implies significance at 0.05 level
* implies significance at 0.10 level
ns implies not significant

Table 5.12: Estimates of weather supply functions of area under rapeseed and mustard in Amritsar

Weather Variable(s)	Intercept	Estimated value of Regression Coefficient on			$(\overline{R})^2$	Durbin-watson Statistic
		W	W^2	T		
Rainfall	23.004	0.232 ns (1.392)	–0.010† (2.397)	0.215 ns (0.540)	0.079	0.652
Maximum Temperature	966.986	–26.301‡ (20.343)	0.385 ns (1.429)	0.289 ns (0.788)	0.083	0.935
Minimum Temperature	183.609	–15.651‡ (13.075)	0.383 ns (1.736)	0.166 ns (0.405)	–0.145	0.935
Mean Temperature	152.884	–7.617‡ (10.018)	0.148 ns (0.354)	0.240 ns (0.633)	0.021	0.878
Relative Humidity	–757.872	–24.625‡ (47.181)	0.193‡ (49.599)	0.460 ns (1.251)	0.081	0.846

‡ implies significance at 0.01 level
† implies significance at 0.05 level
* implies significance at 0.10 level
ns implies not significant

Cotton Desi

Time trend variable—a proxy technology turned out to be negative and significant in both the districts (significant at 0.01 level in Ferozepur and 0.05 level in Bathinda district). Moreover, rainfall has a non significantly negative effect on the area under cotton desi in both the districts of the study area. However, the excess of rainfall has a positive and significant impact in Ferozepur and non significant in Bathinda district. Under this relationship, area under cotton desi changes from one point to another with an optimum at 7.38 cm and 4.21 cm in Bathinda and Ferozepur districts respectively. The excess rain has a positive impact because both these districts are drought prone areas and cotton desi is grown only under rainfed conditions. Furthermore, the most beneficial effect has been noticed for May rains—pre sowing month rains in Bathinda district.

Sugarcane

Pre sowing rainfall has a non significant but negative effect on the area under sugarcane crop, because the crop is sown under assured irrigation and after a heavy preparatory irrigation. The pronounced beneficial effect has been noticed for March rains which is the most suitable time for planting sugarcane. However, the effect was pronounced detrimental for the February rains because the crop is cultivated quickly when the soil comes into *vattar*—right stage of moisture till it is reduced to a fine tilth and the crop is generally cultivated early in the study area.

Rapeseed and Mustard

Rainfall has a non significant but positive effect on the area under rapeseed and mustard crop because the crop is mainly grown under rainfed conditions and that too on the marginal land. However, the excess of rainfall has significantly detrimental effect on the area under the crop because of low water moisture requirements and farmers will shift to wheat crop. Under this relationship, area under rapeseed and mustard changes from one point to another with an optimum at 11.6 cm. The effect was pronounced beneficially for September rains, that is, during the optimum sowing period. However, pre sowing month (August) has detrimental effect on the area under rapeseed and mustard because August showers leave much soil moisture which is conducive for wheat sowing in the study area.

EFFECT OF TEMPERATURE

Air temperature is a decisive factor in plant growth, especially in the sub-tropics. It is closely related with the incoming solar radiation and the energy available for the growth of plants. Photosynthesis in plants is directly related with air temperature upto a certain limits. Each crop is known to have its own optimum (maximum and minimum) temperature conditions. Temperature affects pollination, fertilization and ripening of fruits. Transpiration rates are also closely related with the air temperature. High temperature (mid day) increases the saturation deficits of plants which may also have a damaging effect on the crop during the period of water stress. Stomatae apertures are reduced and ultimately closed and transpiration is checked. It can devernalise the cryophytes, especially the buds of the sun exposed deciduous trees. Coupled with high humidity, it favours the development of many plant diseases. It also affects plant metabolism.

High night temperature increases respiration. It favours the growth of shoots and leaves at the cost of roots, stolons, cambium and fruits. It governs the distribution of photosynthates among different organs of the plants favouring those which are generally not useful. It also affects plants metabolism and accelerates the development of non cryophytes. On the other hand, many crop plants are injured and even killed when night temperature is very low. Tender leaves and flowers are very sensitive to low temperature and frost. Plants which are rapidly growing and flowering are easily killed. Low temperature interferes with the respiration of plants. If low temperature coincides with wet soil, it results in the accumulation of harmful products in the plant cells. Frost also interferes with plant metabolism.

Like rainfall, the effect of temperature on the yield and area of the selected crops in the study area are examined using model 5.19. The results are discussed under two sub sections, namely, the Yield Response and the Area Response.

YIELD RESPONSE

The resulting estimates of the effect of temperature—maximum, minimum and mean, on the yield of selected crops in the study area are presented in Table 5.1 through 5.6. These are discussed crop wise as below:

Paddy

All the measures of temperature, that is, maximum, minimum and mean, had significant beneficent effect on the yield of paddy in Ludhiana district, though the effect was not-significant in case of minimum temperature. The excess of all the measures of temperature had a detrimental effect on the paddy yield because the higher temperature adversely affect not only pollen shedding but also grain filling. Under this relationship yield of paddy changes from one point to another with an optimum at 36.91°C for maximum; 30.81°C for minimum and 33.76°C for mean temperature. The beneficial effect of temperature on paddy yield was noticed throughout the growing season because rice cultivation is conditioned by temperature parameters at different phases of growth. The critical mean temperature for flowering and fertilization changes from 20 to 30°C whereas during ripening the range is from 25 to 35°C. Broadly speaking, for vigorous vegetative growth, moderately high temperature are required whereas for panicle initiation slightly lower temperature would be ideal.

Contrary to this, in Amritsar, all the measures of temperature—maximum, minimum and mean, had detrimental effect on the yield of paddy, though the effect of mean temperature was not significant. This detrimental effect of temperature was noticed throughout the growing season of the crop except the minimum temperature. The minimum temperature of August and September had a beneficial effect on the paddy yield. Under this relationship, yield of paddy in Amritsar changes from one point to another with an optimum at 40.11°C; for maximum temperature; 20.87°C for minimum temperature and 36.03°C for mean temperature.

Wheat

The maximum temperature had a significantly beneficial effect on the yield of wheat in both the selected districts, namely, Ludhiana and Amritsar, though the excess of maximum temperature had detrimental effect on wheat yield. Under this relationship, yield of wheat changes from one point to another with an optimum at 32.52°C in Ludhiana and 34.97°C in Amritsar. This beneficial effect of maximum temperature was pronounced in the month of March, that is, at the ripening stage.

On the otherhand, minimum and mean temperature had a significantly detrimental effect on wheat yield, the excess of these had a beneficial effect, though not significant. Under this relationship, yield of wheat changes from one point to another with an optimum at 8.53°C and 11.43°C of minimum temperature in Ludhiana and Amritsar respectively; and at 19.25°C and 16.57°C for mean temperature for Ludhiana and Amritsar respectively; and at 19.25°C and 16.57°C for mean temperature for Ludhiana and Amritsar respectively. The pronounced detrimental effect of mean temperature was noticed in the month of January, that is, at flowering stage in both the districts while that of minimum temperature in the month of March, that is, at the germination stage in Ludhiana, though the effect was detrimental for most of the growing season.

Cotton American

The effect of all the measures of temperature, maximum; minimum and mean, in determining the yield of cotton american appears to be negligible in both the selected districts, namely, Ferozepur and Bathinda of the study area. None of the coefficients were significant. However, the estimated coefficient carried negative sign for maximum and mean temperature in

Ferozepur and only mean temperature in Bathinda district. Under this relationship, yield of cotton american changes from one point to another with an optimum at 29.62°C, 18.35°C and 19.41°C for maximum, minimum and mean temperature in Ferozepur district. The corresponding figures of temperature for Bathinda district are 32.77°C and 23.31°C and 25.59°C. However, the pronounced beneficial effect of temperature was noticed in the month of May in Ferozepur and November in Bathinda; of minimum temperature was in the month of July in Bathinda and September in Ferozepur while that of mean temperature was in the month of June for Ferozepur and November for Bathinda.

Cotton Desi

The effects of maximum, minimum and mean temperatures on the yield of cotton desi is similar to that of cotton American in both the districts of the study area. However, the mean temperature was noticed to be having a beneficial effect on the yield of cotton desi, though its excess had non significant detrimental effect. Under this relationship, yield of cotton desi changes from one point to another with an optimum at 25.31°C, 19.95°C and 23.21°C for maximum, minimum and mean temperature in Ferozepur district. The corresponding temperatures for Bathinda district are 35.41°C, 22.83°C and 26.38°C. However, the pronounced beneficial effect of maximum and minimum temperature was noticed in the month of June while that of mean temperature for August in Bathinda district. Likewise, the pronounced beneficial effect maximum, minimum and mean temperatures was noticed in the month of May, September and August respectively in Ferozepur district of the study area.

Sugarcane

All the measures of temperatures—maximum; minimum and mean, had a beneficial effect in determining the yield of sugarcane in the study area, though their excesses had non significant detrimental effect. The beneficial effect was because sugarcane crop is able to grow over a prolonged season even under warm humid conditions, unless terminated by flowering. However, its height is strongly influenced by the age of the crop and season. Temperature above 50°C arrest its growth; and below 20°C slow it down markedly and severe frost proves fatal. Under this relationship, yield of sugarcane changes from one point to another with an optimum at 48.69°C for maximum temperature, 24.73°C for minimum temperature and 30.09°C for mean temperature. Furthermore, the pronounced beneficial effect of maximum temperature was observed during the month of September followed by March while that of minimum temperature during May and of mean temperature during the month of October followed by March.

Rapeseed and Mustard

All the measures of temperature had a detrimental effect on the yield of rapeseed and mustard significantly except for minimum temperature where the effect was non significant. This detrimental effect is in accordance with the fact that the crop require relatively cool temperature for its satisfactory growth. Furthermore, the excess of these measures had non significant beneficial effect on the crop yield. Under this relationship, crop yield changes from one point to another with an optimum at 37.06°C for maximum temperature; 24.73°C for minimum temperature and 30.09°C for mean temperature. Furthermore, pronounced detrimental effect of temperature was noticed for the month of October, though the same was noticed for most of the growing season.

AREA RESPONSE

The resulting estimates of maximum, minimum and mean temperature effecting area under selected crops in the study area are presented in Table 5.7 through 5.12. These are discussed crop wise as follows:

Paddy

All the measures of planting season temperature had a beneficial effect on the area under the crop in Ludhiana district, though their excesses had non significantly detrimental effect. Under this relationship, area under paddy changes from one point to another with an optimum at 42.61°C for maximum temperature; 23.36°C for minimum temperature and 30.89°C for mean temperature. Furthermore, the pronounced beneficial effect of all these measures of temperature was noticed in the month of July when most of the transplanting is done.

In Amritsar, the effect of all the measures of temperature on the area under paddy is not much different from that of Ludhiana. The only difference is that in Amritsar, maximum temperature had a detrimental effect on the area under paddy. Under this relationship, area under paddy changes from one point to another with an optimum at 43.93°C for maximum temperature; 27.04°C for minimum temperature and 30.27°C for mean temperature. Furthermore, the pronounced beneficial effect of all the measures of temperature was noticed in the month of May and pronounced detrimental in July because in Amritsar most of the paddy planting is done in the month of May end or early June.

Wheat

All the measures of temperature had a significant detrimental effect on the area of wheat crop in both the selected districts, namely, Ludhiana and Amritsar, except minimum temperature in Ludhiana, where the effect was non significant. Their excesses, however, did not significantly effect the wheat area sown. Under this relationship, area under wheat changes from one point to another with an optimum at 32.46°C for maximum temperature, 21.63°C for minimum temperature and 23.48°C for mean temperature in Ludhiana district. The corresponding optimum temperature are 30.70°C, 19.88°C and 21.95°C for Amritsar district. Furthermore, the pronounced detrimental effect of all the measures of temperature was noticed in the month of November in both the districts under the study with the only expression of maximum temperature in Amritsar. Its effect was pronounced detrimental in the month of October because in Amritsar farmers prefer early sowing of wheat.

Cotton American

All the measures of temperature during sowing period had a detrimental effect on the area under cotton american in both the districts, namely, Ferozepur and Bathinda. However, their excesses had a beneficial effect on area under cotton american. Under this relationship, area under cotton american changes from one point to another with an optimum at 24.34°C for maximum temperature; 15.56°C for minimum temperature and 17.94°C for mean temperature in Bathinda district. The corresponding optimum temperature measures for Ferozepur are 24.86°C, 12.33°C and 21.92°C. Furthermore the pronounced detrimental effect of all the measures temperature was noticed in the month of March in both the selected districts.

Cotton Desi

Contrary to the result of cotton american, all the measure of temperature had a beneficial effect on the area under cotton desi in both the selected districts of the study area. Furthermore, their excesses had a detrimental effect on the cropped area. Under this relationship, area under cotton desi changes from one point to another with an optimum at 37.74°C for maximum temperature; 11.47°C for minimum temperature and 32.99°C for mean temperature in Ferozepur district. The corresponding optimum measure of temperature are 22.21°C, 10.86°C and 15.37°C for Bathinda district. Furthermore, the pronounced beneficial effect of all the measure of temperature was noticed in the month of March in both the selected districts in the study area.

Sugarcane

All the measures of temperature, namely, maximum, minimum and mean temperature had a significantly beneficial effect on the area under sugarcane in the study area. Their excesses, however, did not play any significant role. Under this relationship, area under sugarcane changes from one point to another with an optimum at 32.88°C for maximum temperature; 20.75°C for minimum temperature; and 23.08°C for mean temperature in the study area. Furthermore, the pronounced beneficial effect of all the measures of temperature was noticed in the months of March, though the effect was beneficial through-out the planting period.

Rapeseed and Mustard

All the measures of temperature, namely, maximum, minimum and mean had a detrimental effect on the area under rapeseed and mustard in the study area. This detrimental effect was most pronounced in the month of September. This could be due to the fact that if the temperature did not go down to the desired level the farmers may shift to the other alternative crop, that is, wheat. Their excesses, however, did not play any significant role in determining area under rapeseed and mustard. Under this relationship, area under the crop changes from one point to another with an optimum at 34.16°C for maximum temperature; 20.43°C for minimum temperature and 25.73°C for mean temperature.

EFFECT OF HEAT MOISTURE STRESS

To study the combined effect of temperature and rainfall, another climatic variable, namely, synergistic heat moisture stress, represented by a dummy variable, as discussed earlier, was constructed. The effect of synergistic heat moisture stress variable on the yield and area under the selected crops in the study area was examined through Model 5.19. Its effect did not turn out to be significant in any of the crop with the only exception of Paddy, both on the area and the resultant yield. The resulting estimates of the paddy crop are presented in Table 5.1 and 5.7 respectively for the yield and area under paddy.

The resulting estimates reveals that the synergistic heat moisture stress variable had a detrimental effect on the area and the resultant yield of paddy crop in both the districts, namely, Ludhiana and Amritsar of the study area. However, its excess had no significant effect, though the regression coefficient carried positive sign. This detrimental effect had been noticed through out the sowing season, though it was pronounced detrimental for the month of June on the area under paddy for both the districts under considerations. Likewise, the detrimental effect on the yield of paddy crop was noticed through out the growing season of the crop, though it was pronounced detrimental for the month of July and August. Under this relationship, area under paddy changes from one point to another with an optimum at 190 and 230 respectively for Ludhiana and Amritsar. Likewise, yield of paddy crop changes from one point to another with an optimum at 130 and 118 respectively for Ludhiana and Amritsar district of the study area.

EFFECT OF HUMIDITY

Humidity in the air is of considerable importance in the growth and development of plants. It affects the intensity and the quality of solar radiation reaching the crop and also evapotranspiration rates from the cropped areas. There are certain plants which directly utilize the water vapours in the atmosphere without requiring the change in its state into a liquid form. It is also a source of soil moisture. High humidity reduces the saturation deficit of the plant surfaces and thus reduces transpiration loads of the plant but it favours many plant diseases and pests, otherwise. It increases the growth of shoots and leaves at the expense of fruits, stolons, bulbs and cambium. On the otherhand, low atmospheric humidity increases the saturation deficit and accelerates transpiration. When transpiration is high, water that reaches the leaves is not sufficient for them even when there is no dearth of moisture in the soil. Due to this excessive transpiration, stomatae close and photosynthesis is interrupted. When saturation deficit is extreme, most of the plants wilt and some are even killed because of physiological drought.

The effect of relative humidity on area and yield of the selected crops of the study sample are examined using Model 5.19. The result are discussed under two sub headings, namely, the yield response and the area response.

YIELD RESPONSE

The resulting estimates of the effect of relative mean humidity (RMH) on the yield of the selected crops in the study area are presented in Table 5.1 and discussed, as earlier, crop wise as below:

Paddy

The effect of RMH on the yield of paddy had been noticed to be beneficial for both the districts of the study area, namely, Ludhiana and Amritsar. However, the excess of RMH had signification detrimental effect on the paddy yield, though the effect was non significant for Amritsar. Under this RMH-yield relationship, yield of paddy changes from one point to another with an optimum at 38.66 per cent for Ludhiana and 48.10 per cent for Amritsar. In Ludhiana, the pronounced beneficial effects on yield was noticed for the months of September, August followed by October; though in general, the effect was favourable for whole of the crop season. On the other hand, the effect of RMH on the yield of paddy was mixed in Amritsar. Beneficial effect was pronounced for the month of September followed by June and for the month of October, the effect was pronounced detrimental because of early sowing practices.

Wheat

The effect of RMH on the yield of wheat was noticed to be significantly beneficial, though the excess of RMH had a detrimental effect, in both the selected districts, namely, Ludhiana and Amritsar. Under this relationship, the yield of wheat changes from one point to another with an optimum at 37.58 per cent for Ludhiana and 54.41 per cent for Amritsar. The pronounced beneficial effect of RMH was noticed for the month of November followed by October; though the effect was beneficial through out the crop season in Amritsar. However in Ludhiana, RMH had a mixed effect on the yield of wheat. It was beneficial for the month of October, March and April while detrimental for rest of the crop season, that is, from November through February. However, the pronounced detrimental effect was noticed for the RMH of January.

Cotton American

The effect of RMH on the yield of cotton american was noticed to be beneficial for both the selected districts, namely, Ferozepur and Bathinda. However, the excess of RMH had a detrimental effect on the yield. Under this relationship, yield of cotton american changes from one point to another with an optimum at 51.32 per cent for Bathinda and 56.47 per cent for Ferozepur. Furthermore, the effect of RMH on the yield of cotton american was a mixed phenomena in both the study districts. The effect was beneficial for the months of July through September while it was other way around for rest of the growing season. The effect was pronounced beneficial for RMH of July in Bathinda and August in Ferozepur and detrimental in April and May in both the districts under considerations.

Cotton Desi

The effect of RMH on the yield of cotton desi was noticed to be beneficial for Ferozepur and detrimental for Bathinda district of the study area. Further the excess of RMH had a detrimental effect in Ferozepur and beneficial effect in Bathinda district. Under this relationship, the yield of cotton desi changes from one point to another with an optimum at 49.72 per cent in Ferozepur and 47.95 per cent in Bathinda district. Relative mean humidity during the growing season, however, had a mixed effect on the yield of cotton desi. The ef fect was pronounced detrimental in the months of May followed by April in both the selected districts and pronounced beneficial in the month of September followed by July in Bathinda and August followed by October in Ferozepur.

Sugarcane

The effect of RMH on the yield of sugarcane was noticed to be beneficial in the study area, that is Patiala district, though the excess of RMH had a detrimental effect. Under this relationship, yield of sugarcane changes from one point to another with an optimum at 51.82 per cent. Furthermore, the effect of RMH on the yield of sugarcane was beneficial throughout the growing crop season. Pronounced beneficial effect, however, was noticed in the decreasing order for the month of April, October, November and September.

Rapeseed and Mustard

The effect of RMH on the seed yield of rapeseed and mustard was observed to be beneficial, though the excess of RMH had a detrimental effect. Under this relationship, yield of rapeseed and mustard change from one point to another with an optimum at 58.23 per cent. Furthermore, the effect of RMH was detrimental during the early stages of growth while beneficial at the later stages of growth. The effect was pronounced beneficial for RMH of November.

AREA RESPONSE

The resulting estimates of the effect of relative mean humidity (RMH) on area under selected crops in the study area are presented in Table 5.7 through 5.12 and discussed crop wise as below:

Paddy

Relative mean humidity had a beneficial effect on the area under paddy crop in both the selected districts, namely, Ludhiana and Amritsar of the study area. However, the excess of RMH had a detrimental effect on the area under paddy crop. Under this relationship, area under paddy changes from one point to another with an optimum at 49.00 per cent in Ludhiana and 62.97 per cent in Amritsar. Furthermore, the effect was most beneficial during the month of July followed by June in Ludhiana and May followed by June in Amritsar of the study area because majority of the crop was transplanted during these months.

Wheat

Like paddy, area under wheat crop was also positively related with relative mean humidity, though the excess of RMH had adverse effect in both the selected districts. Under this relationship, area under wheat changes from one point to another with an optimum at 70.31 per cent in Amritsar and 62 per cent in Ludhiana. Furthermore, pronounced beneficial effect of RMH was noticed for the month of November in both the selected districts. During this month, majority of the wheat sowing was completed.

Cotton American

Relative mean humidity had a beneficial effect on the area under cotton american in Bathinda district while the effect was detrimental in Ferozepur. The excess, however, had a detrimental and beneficial effect respectively in Bathinda and Ferozepur. Under this relationship, area under cotton american changes from one point to another with an optimum at 79.77 per cent in Bathinda and 75.14 per cent in Ferozepur. The effect was most beneficial for the month of March in both the districts and pronounced detrimental for April in Ferozepur only.

Cotton Desi

Contrary to cotton american, relative mean humidity had a detrimental effect on the area under cotton desi in Bathinda (its excess, however, had a beneficial effect) and beneficial effect in Ferozepur, while its excess had adverse effect. Under this relationship, area under cotton desi changes from one point to another with an optimum at 69.56 per cent in Bathinda and 72.86 per cent in Ferozepur. The pronounced detrimental effect was noticed in the month of March followed by April and May, that is, through-out the crop sowing season in

Bathinda district while the effect was detrimental only in the month of March in Ferozepur. The effect was pronounced beneficial in Ferozepur in the month of May followed by April.

Sugarcane

Relative mean humidity had a detrimental effect on the area under sugarcane in Patiala while its excess had a beneficial effect. Under this relationship, area under sugarcane changes from one point to another with an optimum at 58.07 per cent. Furthermore, the detrimental effect was noticed throughout the crop sowing season, that is, from January to March and the degree of effect also increase with the advancement of period.

Rapeseed and Mustard

Relative mean humidity had a detrimental effect on the area under rapeseed and mustard in Amritsar while its excess had a beneficial effect. Under this relationship, area under rapeseed and mustard changes from one point to another with an optimum at 63.53 per cent. Furthermore, the detrimental effect was noticed throughout the crop sowing season, that is, August and September. However, the effect was pronounced detrimental in August.

CONCLUDING REMARKS

These resulting estimates of the analysis, discussed in the foregoing pages, reveals that the fitted models explain a major portion of the variation, as given by the high value of $\overline{R}^2$ in the dependent variable, namely, area and yield of the crop under consideration. However, the high value of $\overline{R}^2$ need to be interpreted with a caution. The higher value of $\overline{R}^2$ is obviously due to the highly significant time trend variable which is responsible for explaining most of the variations in the dependent variable.

Although thc crop weather relationship appear so simple to visualise, their empirical estimation, after taking into account the various plant and weather characteristic, is a real tough job, specially in the absence of availability of comparable time series data on various factors affecting crop production. In view of the various constraints, already discussed, the present study has many shortcomings and is based on a number of assumptions. The use of monthly weather records rather than daily or weekly data is a severe limitation. In addition, the use of time trend variable as a surrogate for all non weather factors affecting cropped area and the resultant yield is subject to criticism. Moreover, the study has considered only three of the paraphernalia of the weather factors which affect the crop. The yield of crops, as reported in the publication, is arrived at by averaging over different heterogeneous conditions. The weather effects, thus, gets obscured under such averaging condition. Nevertheless within the data constraints, the study does brought out the role of weather factors in determining the area and yield of the crops and also indicates the need for continuing research and the importance of collection of data under various homogenous conditions. Moreover among the different weather factors considered, rainfall has come out to be the most important factor affecting the area and its resultant yield of the crops studied. At one stage, it has a beneficial effect while at the other detrimental. The results are important in view of the fact that almost the entire area under reference is irrigated. The significance of rainfall in affecting the area sown under the crops and its resultant yield in a predominantly irrigated conditions imply that the available irrigation is not sufficient enough to overcome the deficiency of rainfall and that available irrigation supplements rainfall and is not a substitute. This the general belief that technological innovations such as irrigation can ensure the freedom from the clutches of the vagaries of weather is a myth.

REFERENCES

1. For exhaustive review of the work done on crop-weather relationship, please see, A. Vaidyanathan, "Influence of weather on crop yield: A Review of Agro-Meteorologists Research," *Indian Journal of Agricultural Economics*, Vol. 35, No. 2, April–June 1980.

2. Bernard Oury, "Allowing for Weather in Crop Production Model Building," *Journal of Farm Economics*, Vol. 47, May 1965; and Lawrence H. Shaw, "The Effect of Weather on Agriculture Output: A Look at the Methodology," *Journal of Farm Economics*, Vol. 46, 1964.
3. T.P. Abraham, "Isolation of Effects of Weather on Productivity including other Risks such as Damages by Pest and Diseases," *Journal of Indian Society of Agricultural Statistics*, vol. 17, No. 2, 1965.
4. Harold H. Mann, Rainfall and Famine—A Study of Rainfall in the Bombay Deccan:1865–1938, *Indian Society of Agricultural Economics*, Bombay, 1955.
5. R.J. Kalamkar and V. Satakopan, The Influence of Rainfall Distribution on the Cotton Yields at Government Experimental Farms at Akola and Jalgaon, *Indian Journal of A gricultural Sciences*, Vol. 10, 1940.
6. N.V. Kanitkar, *Dry Farming in India*, Indian Council of Agricultural Research, New Delhi, 1960.
7. L.A. Ramdas, *Crops and Weather in India*, ICAR Monograph, New Delhi, 1960.
8. A.K. Malik, "Effect of Weather on Crop Growth and Yield at Government Experimental Farm at Dharwar," *Journal of Biological Sciences*, Vol. 1, No. 1, 1958.
9. M. Ezekiel and Karl A. Fox, *Method of Correlation and Regression Analysis*, John Wiley and Sons, New York, 1965.
10. R.A. Fisher, "The Influence of Rainfall on the Yield of Wheat at Rothamstead," *Philosophiczal Transactions of the Royal Society of London*, Series B, No. 213, 1924.
11. Only two studies have applied this techniques. These are: P.S. Sreenivasan, "Influence of Rainfall on the Yield of Cotton in Madhya Pradesh," *Indian Journal of Agricultural Sciences*, Vol. 43, No. 9, September 1973; and S.K. Sinha and J.R. Banerjee, "Influence of Rainfall, Humidity, Sunshine, Maximum and Minimum Temperatures on the Yield of Cotton at Coimbatore," *Indian Journal of Meteorology, Hydrology and Geophysics*, Vol. 26, No. 4, 1975.
12. W.A. Hendricks and J.C. Scholl, Techniques in Measuring Joint Effect of Temperature and Precipitation on Crop Yield, North Carolina Agricultural Experiment Station, Technical Bulletin No. 74, 1943.
13. Murray Brown defines technological changes in terms of abstract technology which have four characteristics, viz., (i) technical efficiency; (2) returns to scale; (3) capital intensity, and (4) substitution of inputs. The first two represents neutral technological change, while the latter two are non neutral technological substitution between resources, that is, relative marginal productivities remains constant. See Murray Brown, On the Theory and Measurement of Technological Change, Cambridge University Press, Cambridge, 1968.
14. G.S. Kainth and R.S. Bawa, *Economic Development and Structural Changes—An Empirical Investigation*, Inter India Publications, New Delhi, 1985
15. The assumption does not held true strictly. However, as it is practically impossible to isolate the weather-technology interaction the assumption can be treated as valid within limits.

16. This variable serves as a proxy for the impact of technology like better irrigation and drainages; fertilizer, hybrid seeds, control of pests and diseases, improved agricultural practices and better management etc.
17. R.L. Pitter, "The Effect of Weather and Technology on Wheat Yield in Oregon," *Agricultural Meteorology*, Vol. 18, 1977.

Annex Table 5.1
Estimated value of the weights attached to different weather variables and yield of Paddy

(Month wise)

Weather Variable(s)	June	July	August	September	October
			Ludhiana		
Rainfall	0.312	0.0393	0.3573	0.1234	0.1534
Maximum Temperature	0.1023	0.0602	0.0131	0.1561	0.1522
Minimum Temperature	0.3420	0.1600	0.1359	0.0796	0.1169
Mean Temperature	0.0675	0.0666	0.0527	0.0034	0.0425
Relative Humidity	0.2753	0.1850	0.0765	0.1638	0.1064
Heat Moisture	0.6521	-0.2411	-0.2339	0.0317	0.0507
			Amritsar		
Rainfall	–0.0623	–0.0561	0.3323	–0.0135	0.6065
Maximum Temperature	–0.0332	–0.4719	–0.0386	–0.1953	–0.1132
Minimum Temperature	0.1176	0.0042	0.1661	0.4186	0.0576
Mean Temperature	–0.0912	–0.0924	–0.4344	0.4202	–0.0605
Relative Humidity	0.1233	0.0255	0.1334	0.1499	–0.1181
Heat Moisture Stress	0.7503	–0.2331	–0.2331	–0.0297	0.0409

Annex Table 5.2
Estimated value of the weights attached to different weather variables and area under paddy

(Month wise)

Weather Variable(s)	May	June	July
	Ludhiana		
Rainfall	0.3950	0.0067	0.1641
Maximum Temperature	0.4435	0.2174	0.2141
Minimum Temperature	0.1670	0.3144	0.1659
Mean Temperature	0.3324	0.5056	0.3711
Relative Humidity	0.2320	0.1214	0.1190
Heat Moisture Stress	–0.0623	–0.4379	–0.0985
	Amritsar		
Rainfall	0.4269	0.1736	0.2659
Maximum Temperature	–0.1741	0.0485	–0.4668
Minimum Temperature	–0.0463	–0.0442	0.0110
Mean Temperature	0.1295	–0.6804	–0.0688
Relative Humidity	0.0882	0.5406	–0.1015
Heat Moisture Stress	–0.0579	–0.5249	–0.0789

Annex Table 5.3
Estimated value of the weights attached to different weather variables and yield of wheat

(Month wise)

Weather Variable(s)	January	February	March	April	October	November	December
			Ludhiana				
Rainfall	0.3576	0.0896	0.3864	–0.2774	0.550	0.5065	0.0709
Maximum Temperature	–0.1931	–0.0609	0.0764	0.0625	0.1781	0.3068	–0.0832
Minimum Temperature	–0.3626	–0.3012	0.2701	–0.1415	0.1096	–0.3373	–0.0696
Mean Temperature	–0.2050	–0.0273	0.1751	–0.2707	0.1463	0.1693	–0.2797
			Amritsar				
Rainfall	–0.0152	–0.0287	–0.2453	–0.0420	–0.2642	–0.3916	–0.3260
Maximum Temperature	–0.4358	–0.0378	–0.0971	0.0824	–0.1357	0.3736	0.2498
Minimum Temperature	–0.2759	–0.1549	–0.0068	0.2789	0.0313	–0.4343	–0.0346
Mean Temperature	0.0348	0.0597	0.1745	0.1457	0.2625	0.1199	0.0900
Relative Humidity	–0.0989	–0.1188	–0.0944	–0.2771	–0.2247	–0.2122	–0.2834

Annex Table 5.4
Estimated value of the weights attached to different weather variables and area under wheat

(Month wise)

Weather Variable(s)	September	October	November
	Ludhiana		
Rainfall	–0.1863	–0.1599	0.1906
Maximum Temperature	0.3270	0.0100	0.2373
Minimum Temperature	0.1333	0.1710	–0.3989
Mean Temperature	0.2213	0.1594	–0.2088
Relative Humidity	–0.0479	–0.1883	–0.2614
	Amritsar		
Rainfall	0.2633	–0.0626	–0.0403
Maximum Temperature	–0.1385	0.1766	0.3710
Minimum Temperature	0.1638	0.2108	–0.2692
Mean Temperature	0.2042	0.4164	0.2943
Relative Humidity	0.0677	–0.0690	–0.3208

Annex Table 5.6
Estimated value of the weight attached to different weather variables and area under cotton American

(Month wise)

Weather Variable(s)	March	April	May
	Ferozepur		
Rainfall	0.3222	0.1107	0.2867
Maximum Temperature	–0.2174	0.2601	–0.1867
Minimum Temperature	0.0578	0.2893	0.1737
Mean Temperature	–0.0550	0.3377	–0.1280
Relative Humidity	–0.3533	0.2981	0.4745
	Bathinda		
Rainfall	0.0726	–0.2701	0.2692
Maximum Temperature	–0.3925	–0.3946	–0.1388
Minimum Temperature	0.0396	–0.0081	–0.1911
Mean Temperature	–0.1724	0.2491	–0.3776
Relative Humidity	0.2233	0.1154	0.0645

Annex Table 5.5

Estimated value of the weights attached to different variables and yield of cotton American

(Month wise)

Weather Variable(s)	April	May	June	July	August	September	October	November
				Ferozepur				
Rainfall	–0.1533	0.0061	–0.1353	–0.0936	–0.0586	0.2652	–0.0797	0.1149
Maximum Temperature	0.1450	0.5838	–0.0931	–0.1180	0.3180	–0.2322	–0.1380	0.0048
Minimum Temperature	–0.3195	–0.2715	–0.1175	–0.6156	–0.2671	–0.1502	–0.1853	0.0250
Mean Temperature	–0.2147	0.1072	0.1756	–0.274	0.1480	–0.3552	–0.3914	–0.2978
Relative Humidity	0.1062	–0.0664	–0.2046	–0.2764	0.0495	–0.1238	–0.1713	0.1438
				Bathinda				
Rainfall	–0.4317	–0.0808	–0.0595	–0.1880	0.1036	–0.0762	–0.4762	0.2294
Maximum Temperature	0.0126	–0.0961	–0.0262	–0.1929	–0.1245	0.1212	0.3995	–0.2145
Minimum Temperature	–0.2300	–0.0098	0.4394	0.3346	0.1386	0.2300	0.0537	–0.0577
Mean Temperature	–0.1029	–0.0367	–0.0318	–0.0474	0.1904	–0.1464	–0.4300	–0.0729
Relative Humidity	–0.3239	–0.3100	–0.1432	0.0720	0.1137	0.1163	0.1573	–0.0339

Annex Table 5.7

Estimated value of the weights attached to different weather variables and yield of cotton desi

(Month wise)

Weather Variable(s)	April	May	June	July	August	September	October	November
Ferozepur								
Rainfall	–0.2908	–0.1910	–0.0735	–0.1051	–0.0409	0.3506	–0.1244	0.0565
Maximum Temperature	–0.0131	0.0623	–0.2391	0.2581	0.3044	0.1990	–0.0762	0.3516
Minimum Temperature	–0.2472	–0.3291	0.0489	–0.6724	–0.1484	0.0780	0.0247	0.1218
Mean Temperature	–0.2908	–0.1910	–0.0735	–0.1051	–0.0408	0.3506	–0.1244	0.0565
Relative Humidity	–0.2399	–0.4345	–0.2850	0.0257	–0.3427	0.0283	–0.1867	–0.0552
Bathinda								
Rainfall	0.2966	–0.0317	–0.2623	0.2279	0.1712	0.1056	0.1482	0.1348
Maximum Temperature	–0.1266	0.0053	–0.1115	–0.2138	–0.5413	–0.1688	0.0528	–0.1778
Minimum Temperature	–0.2893	0.1970	0.1846	–0.0435	–0.1174	0.2657	–0.0214	0.0103
Mean Temperature	–0.2340	–0.2073	–0.0422	–0.0434	0.4056	–0.0884	–0.3804	0.0211
Relative Humidity	0.0755	–0.0543	–0.0280	0.1373	0.0670	0.961	0.1563	0.1672

Annex Table 5.8
Estimated value of the weight attached to different weather variables and area under cotton desi

(Month wise)

Weather Variable(s)	March	April	May
	Ferozepur		
Rainfall	–0.1416	0.0746	0.1781
Maximum Temperature	0.3364	–0.2152	0.1513
Minimum Temperature	0.2008	–0.2025	0.3367
Mean Temperature	–0.0291	0.1223	–0.0371
Relative Humidity	0.2880	–0.0587	0.0612
	Bhatinda		
Rainfall	–0.1437	–0.1499	–0.2201
Maximum Temperature	0.2502	0.1819	0.1437
Minimum Temperature	0.0811	0.1129	–0.0049
Mean Temperature	0.2487	–0.0616	0.1271
Relative Humidity	0.1526	0.3201	–0.0116

Annex Table 5.10
Estimated value of the weights attached to different weather variables, yield and area under rapeseed and mustard in Amritsar

(Month wise)

Weather Variable(s)	August	Sept.	Oct.	Nov.	Dec.
		Yield			
Rainfall	—	0.0309	0.0967	0.0551	0.1213
Maximum Temperature	—	–0.1188	–0.2523	–0.0701	0.0649
Minimum Temperature	—	0.1222	0.3009	0.4098	0.0562
Mean Temperature	—	0.4195	–0.1539	0.0459	–0.1154
Relative Humidity	—	0.1069	0.2182	0.2988	0.2560
		Area			
Rainfall	–0.2611	–0.2511			
Max. Temperature	0.2550	0.1635			
Min. Temperature	0.0433	0.0467			
Mean Temperature	–0.1217	–0.1230			
Relative Humidity	–0.2134	–0.0631			

Annex Table 5.9

Estimated value of the weights attached to different weather variables, yield and area under sugarcane in Patiala

(Month wise)

Weather	Jan	Feb	March	April	May	June	July	August	Sept	Oct	Nov	Dec
Yield												
Rainfall	—	–0.5285	0.2998	0.2995	–0.0138	0.3805	0.2528	0.1943	0.2115	0.1261	0.2979	0.2941
Maximum Temperature		–0.1850	–0.4331	–0.3353	–0.3557	–0.0209	0.1285	0.1375	–0.0644	0.0868	0.1634	–0.2843
Minimum Temperature		0.1713	–0.0084	–0.5423	–0.4641	0.3318	0.2033	–0.2402	0.0386	0.2465	0.0094	–0.0099
Mean Temperature		0.1913	–0.3238	–0.5057	–0.2768	0.1957	0.3492	–0.2578	0.0783	0.0358	0.0789	–0.1298
Relative Humidity		–0.0689	–0.1065	0.1332	0.0424	–0.5321	–0.2906	0.4818	0.3921	0.1465	0.0934	0.2574
Area												
Rainfall	–0.1898	–0.3237	0.2266									
Maximum Temperature	–0.3902	–0.1672	–0.0488									
Minimum Temperature	–0.1063	–0.4205	0.3402									
Mean Temperature	–0.3558	–0.3923	0.0682									
Relative HumidityHumidity	0.1299	0.1322	0.1739									

Annex Table 5.11
Zero order correlation matrix of selected weather variables and yield of paddy

(Month wise)

Weather Factor	June	July	August	Sept	Oct
		Amritsar			
Rainfall	0.1049	0.2951	0.2127	0.3141	0.0592
Maximum Temperature	–0.0365	–0.3326	–0.1709	–0.2001	–0.0540
Minimum Temperature	–0.1604	–0.1778	0.2807	0.1214	–0.2568
Mean Temperature	–0.4650	–0.1776	–0.4013	–0.0075	0.0699
Relative Humidity	0.4599	–0.2364	–0.0181	–0.6011	–0.2826
Heat Moisture Stress	–0.3612	–0.1150	–0.1150	–0.3223	–0.2269
		Ludhiana			
Rainfall	0.1780	0.2330	0.4240	0.2090	0.2340
Maximum Temperature	0.3726	0.1538	0.5752	0.6429	0.4669
Minimum Temperature	0.2047	0.0027	0.2375	0.1218	0.3121
Mean Temperature	0.4503	0.4715	0.5943	0.6819	0.5890
Relative Humidity	0.3680	0.2701	0.3604	0.4119	0.7230
Heat Moisture Stress	–0.3175	–0.1192	–0.1192	–0.3346	–0.2768

Annex Table 5.12
Zero order correlation matrix of selected weather variables and area under paddy

(Month wise)

Weather Factor	May	June	July
	Amritsar		
Rainfall	0.2623	0.0763	0.1010
Maximum Temperature	0.3181	–0.2200	–0.2556
Minimum Temperature	0.2400	–0.1375	–0.2791
Mean Temperature	0.2316	–0.3891	–0.3436
Relative Humidity	0.4996	0.2386	–0.0382
Heat Moisture Stress	–0.2513	–0.3549	–0.2268
	Ludhiana		
Rainfall	0.3252	0.1867	0.2398
Maximum Temperature	0.2884	0.2078	0.1813
Minimum Temperature	0.0063	0.1047	0.2493
Mean Temperature	0.0800	0.2880	0.5330
Relative Humidity	0.1870	0.0090	0.2040
Heat Moisture Stress	–0.2734	–0.4189	–0.3123

Annex Table 5.13

Zero order correlation matrix of selected weather variables and yield of wheat

(Month wise)

Weather Factor	Jan	Feb	March	April	Oct	Nov	Dec
			Amritsar				
Rainfall	0.1379	0.0385	–0.2868	0.2436	–0.1289	0.0748	–0.4363
Maximum Temperature	0.2665	–0.1870	–0.3738	–0.0960	–0.1654	0.1501	0.2993
Minimum Temperature	0.0280	0.0511	–0.2962	0.2579	–0.0671	–0.0831	0.1812
Mean Temperature	0.2680	–0.1671	–0.0671	–0.0025	–0.0122	0.1010	0.1899
Relative Humidity	0.1765	0.1465	0.1213	0.0804	0.2047	0.3716	0.0341
			Ludhiana				
Rainfall	–0.1998	0.1165	–0.0925	0.0623	–0.0343	–0.1582	–0.6360
Maximum Temperature	0.0578	–0.2662	–0.8720	–0.0219	0.1814	–0.2257	–0.3306
Minimum Temperature	0.1183	–0.0671	–0.0957	0.0268	–0.2099	–0.0281	0.0336
Mean Temperature	–0.6102	–0.3836	–0.0312	–0.0187	–0.2994	0.0159	0.0605
Relative Humidity	–0.3117	–0.0028	0.1825	0.0448	0.1869	–0.0293	–0.1486

Annex Table 5.14
Zero order correlation matrix of selected weather variables and area under wheat: Month wise

(Month wise)

Weather Factor	September	October	November
	Amritsar		
Rainfall	0.3918	0.2235	0.3128
Maximum Temperature	0.0373	–0.4392	–0.1340
Minimum Temperature	–0.0189	0.0985	–0.1773
Mean Temperature	–0.0748	–0.1383	–0.1497
Relative Humidity	0.2977	0.4807	0.6682
	Ludhiana		
Rainfall	0.2588	0.1686	0.2684
Maximum Temperature	–0.5220	–0.0357	–0.3873
Minimum Temperature	0.0321	–0.2355	0.2753
Mean Temperature	–0.4687	–0.2503	–0.0966
Relative Humidity	–0.0675	0.4259	0.4623

Annex Table 5.16
Zero order correlation matrix of selected weather variables and area under cotton American

(Month wise)

Weather Factor	March	April	May
	Bathinda		
Rainfall	–0.2656	0.4004	–0.2720
Maximum Temperature	–0.0920	0.0679	0.1811
Minimum Temperature	0.0396	–0.0081	–0.1911
Mean Temperature	–0.5366	–0.0455	0.1156
Relative Humidity	0.6714	0.5879	0.4368
	Ferozepur		
Rainfall	0.3891	–0.2272	0.1068
Maximum Temperature	0.1838	–0.0974	–0.1417
Minimum Temperature	0.2744	0.0394	–0.1603
Mean Temperature	0.3320	0.2661	–0.0826
Relative Humidity	0.6421	–0.6826	–0.5312

Annex Table 5.15
Zero order correlation matrix of selected weather variables and yield of cotton American

(Month wise)

Weather Factor	April	May	June	July	Aug	Sept	Oct	Nov
				Bathinda				
Rainfall	–0.7258	0.0559	0.2801	0.3003	0.2551	–0.2612	–0.7370	0.2055
Maximum Temperature	0.0903	0.1493	0.2163	–0.3042	–0.1319	–0.0281	0.2069	0.2744
Minimum Temperature	0.0994	–0.0588	0.3617	0.4669	0.3435	0.2524	0.0091	–0.1275
Mean Temperature	0.3289	–0.1926	0.0278	0.0065	0.0947	0.0612	0.1318	0.7042
Relative Humidity	–0.5840	–0.5965	–0.0165	0.2252	0.1189	0.2223	0.0651	–0.1590
				Ferozepur				
Rainfall	–0.5951	–0.1291	–0.5937	–0.6131	–0.0548	–0.2767	–0.1129	–0.1206
Maximum Temperature	0.3878	0.5222	0.1183	0.2456	0.1486	–0.1086	0.1911	0.4443
Minimum Temperature	0.1289	0.0129	0.1629	–0.3819	0.2067	0.4743	0.2508	0.3135
Mean Temperature	0.2673	0.2551	0.4464	0.2027	–0.0376	0.1409	0.0476	0.0340
Relative Humidity	–0.5785	–0.6531	–0.1718	0.0873	–0.4194	0.0202	–0.2504	–0.2087

Annex Table 5.17

Zero order correlation matrix of selected weather variables and yield of cotton desi

(Month wise)

Weather Factor	April	May	June	July	Aug	Sept	Oct	Nov
Bathinda								
Rainfall	0.6376	0.2286	0.0842	0.2322	0.2096	–0.1407	–0.6637	0.1902
Maximum Temperature	–0.0993	0.0568	0.3593	–0.3880	–0.2414	–0.0368	0.4160	0.3292
Minimum Temperature	–0.0560	–0.2330	0.5696	0.5461	0.2988	0.2561	0.2252	–0.0298
Mean Temperature	0.3059	–0.3145	0.1153	0.1174	0.3926	–0.0081	0.0411	–0.1396
Relative Humidity	–0.5103	–0.5770	–0.0913	0.2243	0.0678	0.2372	0.0350	–0.1575
Ferozepur								
Rainfall	0.5494	–0.1614	–0.0865	0.2027	–0.0799	–0.2103	–0.1956	–0.0064
Maximum Temperature	–0.0725	0.2535	0.1509	0.1766	0.2100	–0.1738	0.1489	0.2935
Minimum Temperature	0.0546	–0.0877	0.3375	–0.3448	0.2693	0.4337	0.2969	0.1179
Mean Temperature	–0.1595	–0.0371	–0.0272	–0.0501	0.1855	–0.2750	–0.3618	–0.0666
Relative Humidity	–0.6274	–0.6834	–0.1726	0.0570	–0.5431	–0.1027	–0.3443	–0.2758

Annex Table 5.18
Zero order correlation matrix of selected weather variables and area under cotton desi

(Month wise)

Weather Factor	March	April	May
	Bhatinda		
Rainfall	–0.1771	–0.3258	0.3902
Maximum Temperature	0.2240	0.0065	–0.1186
Minimum Temperature	–0.0198	0.0265	–0.1834
Mean Temperature	0.5658	0.0648	–0.0055
Relative Humidity	–0.6424	–0.4308	–0.3371
	Ferozepur		
Rainfall	–0.0658	–0.1358	0.2041
Maximum Temperature	0.3364	–0.2152	0.1513
Minimum Temperature	0.2626	0.0835	–0.2316
Mean Temperature	0.4843	0.2165	–0.0415
Relative Humidity	–0.9523	0.1526	0.3260

Annex Table 5.19
Zero order correlation matrix of selected weather variables and area under sugarcane in Patiala

(Month wise)

Weather Factor	January	February	March
Rainfall	–0.1898	–0.1978	0.2365
Maximum Temp.	0.2233	–0.1639	0.2801
Minimum Temp.	–0.1609	0.1069	0.2263
Mean Temp.	0.3774	0.4957	0.4167
Relative Humidity	–0.2826	–0.3686	–0.4298

Annex Table 5.20

Zero order correlation matrix of selected weather variables and yield of sugarcane in Patiala

(Month wise)

Weather Factor	Feb	March	April	May	June	July	Aug	Sept	Oct	Nov	Dec
Rainfall	–0.1707	0.5420	0.5397	0.3252	0.6423	0.5882	0.2233	0.3513	0.1863	0.0053	0.0492
Maximum Temperature	–0.0791	–0.3440	–0.2152	–0.2943	–0.6255	0.0208	0.1237	–0.3810	–0.2778	–0.2794	0.0176
Minimum Temperature	–0.1787	0.1471	–0.0257	0.3326	–0.6707	–0.0143	–0.2752	0.1224	0.0652	0.1550	–0.0736
Mean Temperature	–0.5065	–0.5170	–0.3528	0.1062	–0.6028	0.1668	0.0696	–0.0712	–0.6176	–0.2083	–0.1153
Relative Humidity	0.3977	0.2667	0.8134	0.3294	0.6386	0.3927	0.5459	0.6033	0.6534	0.6341	0.5161

Annex Table 5.21
Zero order correlation matrix of selected weather variables and area under rapeseed in Amritsar

(Month-wise)

Weather Factor	August	September
Rainfall	–0.1755	–0.2760
Maximum Temperature	0.2585	0.1678
Minimum Temperature	0.0544	–0.1328
Mean Temperature	–0.0985	–0.4577
Relative Humidity	–0.3066	–0.1752

Annex Table 5.22
Zero order correlation matrix of selected variables and yield of rapeseed in Amritsar

(Month-wise)

Weather Factor	September	October	November	December
Rainfall	0.0051	0.5008	0.0211	0.6695
Maximum Temperature	0.0966	–0.0593	–0.1224	–0.0057
Minimum Temperature	–0.0874	–0.1159	0.1075	0.0939
Mean Temperature	0.0886	–0.3221	–0.1691	–0.1640
Relative Humidity	–0.1014	–0.0192	0.3057	0.1258

Chapter VI

SUPPLY BEHAVIOUR IN AGRICULTURE

Untill recently, agriculture in developing economies was believed to be tradition-ridden and hence non-responsive or perverse to economic stimuli. It was believed that farmers do not allocate resources as guided by prices, marginal costs and returns consideration. This notion of perversity was shaken by the empirical works in the early sixties which shows that the farmers in developing economies not only respond to price changes and varied farm production in the manner as did the farmers in developed economies. But even the magnitude of response in two widely different economies nearly matched. The importance of price response of supply is derived from the fact that a perverse relationship or absence of relationship between changes in prices and supply blunt the policy instrument for state action. The study of the farmers response and the measurement of price elasticities is an important area of research, particularly in the developing counties where technical progress and the expanding irrigation have led to changes in the cropping pattern and shortages in certain key crops.

Typically acreage—a proxy for production, is viewed as a major decision making variable with respect to input and output prices. Discrepancies between planned and the actual production owing to climatic factors which are beyond the farmers control are the main argument to consider land as the best decision making variable.[1] Moreover, the responsiveness of yield to price was generally assumed to be negligible until the late 1970s. A number of studies focusing on yield have emerged since then.[2] There is, however, a mixed evidence regarding the impact of input prices, particularly of fertilizer. Moreover, it is the responsiveness of total output to price which is important to policy makers. Total output is determined by yield and area planted. The latter can be only ignored if it remains constant. It will not remain constant, even if all available land is under cultivation, if there are opportunities to substitute one crop with another. In this study we, therefore, estimate supply function for both acreage and yield; output response is derived as a reduced form.

In most supply function for developing countries surprisingly, rainfall, which is a crucial variable in the determination of acreage and the yield of crops in developing countries, has been treated some what cursorily. Agriculture in many developing countries is still predominantly rain dependent and unless rainfall is introduced into the supply response function carefully, the equation can be mis-specified. This can lead to bias in the computation of elasticities. The source of the problem lies in distinguishing between fluctuation in acreage and yield due to variation in rainfall (which are at times quite large) and due to other variables such as prices and irrigation etc.

One of the aims of this chapter is to remove the mis-specification problem by carefully modelling rainfall. We discard the usual practice of introducing rainfall in a monotonic — either simple linear or logarithmically linear–fashion and replace it with a non-linear specification that recognises the detrimental impact of floods as well as droughts. Apart from improving the specification considerably, this enables us to derive the optimum amount of rainfall required for crops and to see whether or not a crop is becoming increasingly dependent on rain. This has additional implications for agricultural investment policy which are discussed in the penultimate section of the chapter.

THEORETICAL FRAMEWORK

The nature and the extent of a farmer's response to changes in prices, remains a widely debated and controversial issue.

One school of thought holds the view that peasants in the UDCs (under developed countries) are not responsive to changes in the relative prices and lack motivation. The agricultural supply function has a elasticity near zero. B.D. Giles (1951) doubts whether the model of profit maximisation is applicable to agriculture. To quote him:

> "there is not a little evidence to show that, of the output which are technically possible, the farmers will produce a collection very like that which would yield the greatest profit, although, in fact, he may never operate at the point of equilibrium."

The lack of responsiveness of agricultural production to changes in prices is attributed to number of factors which are operative in agricultural production. The historic explanation for the disappointing response of the farmers to price incentive is that farming is a way of life. A great part of the production is meant for self-consumption. Under a subsistence form of agriculture, the individual farmers are not concerned with price and price relationship, as the production is consumed within the household. While it is true that a great part of farm production is consumed by household, and that part of the produce will be price irresponsive, the portion of the area or the produce meant for the market can reasonably be expected to respond to the price changes. On this basis, it can be said that the farm production response to price may be low, but it cannot be absent altogether, as in no country the whole of the produce is consumed by farm producers. The farmers produce certain crops meant purely for the market and it is expected, that in the case of these market crops, the profit motivation must be affective and changes in the relative prices of food crops and the market crops ought to influence the decision of the farmers in allocating the area and other resources to the production of these two types of crops. We can expect that the food crops, which are grown largely for self consumption, will be less responsive to price as compared to the responsiveness of the market crops, but we cannot preclude the role of market mechanism even in the case of food crops, We can only expect a distinction in the degree of responsiveness of these two types of crops.

Another explanation given for the lack of responsiveness of aggregate agricultural production to price changes is that agricultural costs are composed largely of fixed costs. The proportion of farm inputs which the farmers purchase is very small and a large part of the resource come from within the family and the farm. The family labour, livestock, agricultural machinery, farm building and agricultural land assume the character of fixed cost and these fixed factors are not adaptable for use in the non-farm sector, so that they continue to be used in agriculture even if the price fall sharply. Only the variable factors, which constitute small proportion of the total farm inputs, can be discarded, that is, if the price does not cover the variable costs, aggregate agricultural production can be less price elastic. But here too we cannot exclude the role of the price mechanism in the decision of the farmers to use the variable factors, even if they constitute a small proportion of the total farm inputs in the process of agricultural production. If the farmers cannot discard the fixed factors, they are likely to dispense with the variable factors from employment. Thus we can say that the response of agricultural production may be low, but there is no reason why it ought to be absent.

In a developing agricultural system, the production of variable factors in the total farm inputs is low. But as farming begins to be carried out along commercial lines and as the opportunity for employment of farm labour increases in the non farm sector, the importance of fixed factor goes on declining. In the developed system of agriculture, the proportion of the variable factors is high, so that the ability of adjustment of agricultural production to the price change is also high. It is suggested, therefore, that if farms were organised on a corporate basis and if hired labour were used instead of family labour, the producers would have behaved just like industrial producers.

We may now examine another commonly held belief that the agricultural supply curve is backward bending. The explanation given to account for the perversity of the response of aggregate output to price is that the farmer's requirement for money income is fixed, so that they are required to produce and sell more when agricultural price tend to rise. On this presumption, D.R. Khatkhate (1962) and N.A. Khan (1963) have claimed that in India the farmers have a backward bending supply curve. Though their analysis relates to the supply curve of the marketed surplus of agricultural produce, it also applied to aggregate agricultural production. To quote Khan:

> "The economic organisation is such that within limits, a fall in price stimulates supply of farm products, and a rise in prices check(s) it. We get in agriculture what is known as Inverted Supply Curves."

This perversity in the response of aggregate agricultural production is also believed to exist in other countries, at least in the developing countries, where a large part of the produce is meant for self consumption by the farmers. For example, J.H. Boeke (1942) of Indonesia believes in the backward bending supply curve of agricultural production. All these economists who have advanced the case of perversity in the response of the farmers have in their minds, what can be called 'target income'. They argued that the farmers have a fixed money obligation to pay net rent, debt and interest thereon, to purchase cloth and other non farm products of their requirements.

But there seems to be no reason to assume that the money requirement should be fixed. In a dynamic economy the demand for money income is always increasing. In a developing country where the level of per capita income in agriculture is so low that the needs of the bulk of the farmers are always unsatisfied, and the marginal utility of income is not expected to be so low, the negative income effect of the improvement of the agricultural terms of trade would outweigh the positive substitution effect. The demand for money income is high in the agricultural sector, even for acquiring traditional forms of wealth, such as land, gold and other precious metals. In a dynamic economy, the demand for agricultural products is rapidly increasing, providing profitable opportunity for investment in agriculture. The farmer's demand for capital expenditure such as on the creation of irrigation works, fertilizer and land improvement can also be expected to increase.

Besides, the demand for durable consumer goods such as bicycles, sewing machines, transistors, radio, television and so on, will be expected to increase due to the demonstration effect[3] of the urban civilisation, which will induce the farmers to produce and sell more to acquire more money income. It will be wrong to assume that there is no change in the habits of the farmers. The contact with the urban civilisation through the development of transport facilities, electrification and education will cause a disequilibrium in the desired and existing level and pattern of consumption in the rural population and may induce them to produce and sell more.

PARTIAL ADJUSTMENT

The hypothesis of partial adjustment suggests that there may be constraints, both technological as well as institutional, in effecting full adjustments to the desired position within the short time span of one period. Clearly, this is not very relevant for the yield equation but is quite important for the area equation.

The farmers face a number of constraints while making production decisions in response to changes in the economic stimuli. It is very seldom that they are able to make cent per cent adjustments in responding to various economic stimuli, or adjust instantaneously. In an agriculture which is subject to weather uncertainties, or which is under going changes in production technology as is the case in some developing countries, such constraints become still more severe. Under such conditions, the adjustment lag model is considered more appropriate for measuring farmer's response behaviour.

Variation in A_t^* is connected by variations in A_t observed or actual supply by assuming the following relationship between the actual and long run desired level of supply:

$$A_t = A_{t-1} + \lambda(A_t^* - A_{t-1}) \qquad \ldots \quad (1)$$

The current supply then is:

$$A_t = A_{t-1} + \lambda(A_t^* - A_{t-1}) \qquad \ldots \quad (2)$$

λ is the coefficient of adjustment. The forces which cause the difference in the short and long run elasticities of supply will also determine λ. The first equation is a behavioural relationship, stating that the desired acreage under the crop studied depends upon the relative acreage in the preceding year. The second equation is the partial- area- adjustment equation in which λ is the coefficient of adjustment. The value of λ lies between 0 and 1 and is based on Hick's elasticity of expectations. The farmers, it is hypothesised, are able to change the acreage of a crop in any year only to the extent of the fraction λ of the difference between the acreage they will like to sow and the acreage actually sown in the preceding year. λ therefore, is an indication of how fast the farmers are adjusting themselves to their expectations. A value of λ close to zero, would mean that farmers are adjusting themselves to the changing prices, yield etc. very slowly. A value of λ close to unity would mean that the farmers are adjusting to the changing situation almost instantaneously, full adjustment within a single period.

UNCERTAINTY REGARDING PRICES

In the dynamic setting of supply response, it is necessary to pay attention to the theories relating to risks and uncertainty, so also to adjustment over time and fixity of resources, if suspected. For most of the farm commodities, farmers have to commit inputs sometimes before output is realized. Naturally, they have to base their plans on the expected course of prices in the future. It is difficult to observe subjective price expectations of individual farmer for a long period of time. Hence, theoretical formulation of plausible process of average expectations of a group of farmers is necessary. Such theoretical farmers decisions at the micro level and their impact on supply at the macro level over time remains a widely debated and controversial issue. One such postulation originally formulated by Cagan and used extensively in supply response studies by Nerlove and other authors is as follow:

$$P^*_t - P^*_{t-1} = \phi\ (P_{t-1} - P^*_{t-1})\ 0<\phi<1 \quad \ldots \quad (3)$$

that is,

$$P_t^* = \Phi \sum_{i=0} (1-\Phi)^i P_{t-1}$$

where P^*_t is expected normal price.

The above formulation suggests that a long run normal price is a weighted moving average of past actual prices and weights decline as we go back in time. We have two extreme cases, viz,(i) when $\phi = 0$, and (ii) when $\phi = 1$. The first case is to be ruled out as unrealistic because it suggests that actual prices have no effect on expected prices. The second case corresponds to the type of forecast generated by a native model where

$$P_t^* = P_{t-1}$$

INCORPORATING RAINFALL IN SUPPLY RESPONSE FUNCTIONS

Most supply response functions for developing countries include rainfall as an important explanatory variable. Typically, rainfall appears in the area and yield equation in a monotonic fashion. Of the total cultivable area available, farmers have to decide on the proportion to be allocated to one crop or the other. The decision on acreage is determined in parts by anticipated relative prices, since price of output and of fertilizers are uncertain at the time of sowing. The agronomic variable that affect soil moisture content are chiefly water from irrigation (reflected by the proportion of area under irrigation) and rain during the sowing months.

Once the acreage decision is taken, yield is dependent on the price of inputs — which are known — relative to the output price, which is uncertain. Yield is also affected by the proportion of area that is irrigated and the amount of rainfall during the growing period.

If the estimated value of the regression coefficients for rainfall are positive, this implies that acreage and yield go up with the increased rainfall [see Fig.6.1 panel (a)]. This obviously rules out the excessive rain, having an adverse impact on output,which is a common occurrence in developing agriculture where drainage systems are poor.

Instead of this monotonic specification, it clearly would be preferable to have rainfall positively related to yield or acreage upto some optimal point (R^*) after which it reduces yield or acreage. Two alternatives to the monotonic specification are given in panels (b) and (c) of Fig. 6.1 .

In the yield equation, (the corresponding relationship between acreage and rainfall can be derived in an analogous fashion), one alternative to the monotonic specification of rainfall is bell-shaped normal distribution type specification which assumes that too much rainfall is as bad as too little rainfall [panel (b)]. Letting the constant term K subsume the effects of price and irrigation, we can illustrate this as:

$$\frac{O_1}{A_1} = Ke^{-\beta(R_g - R_g^*)^2}, \beta > 0 \qquad \ldots \quad (4)$$

Where R_g* is the optimum rainfall, that is, yield is at maximum point when:

$$R_g = R_g^* \qquad \ldots \quad (5)$$

Taking logarithm on both sides, we get:

$$\log(\frac{O_1}{A_1}) = \text{Constant} - \beta R_g^2 + 2\beta R_g^2 R_g \qquad \ldots \quad (6)$$

Notice that R^*_g, the optimum growing season rainfall can be recovered from the estimated coefficient of (6): it is the negative of the coefficient of R_g divided by twice the coefficient of R^2_g. However, one disadvantage of (4) is that it assumes that the harmful effects of a deviation of rainfall on either side of the optimum are symmetric. To eliminate this and allow for an asymmetry of impacts of shortage and excess we take the clue from the familiar gamma curve [panel (c)] and specify:

$$\frac{O_1}{A_1} = Ke^{-cR_g} R_g^p . c > 0; p > 0 \qquad \ldots \quad (7)$$

This skewed specification implies that the detrimental impact of droughts is greater than that of floods, which seems

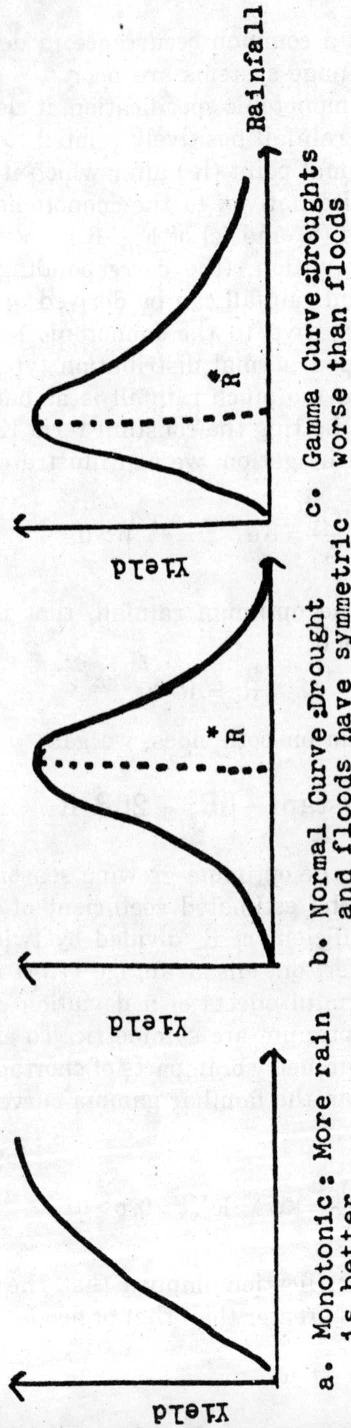

Fig. 6.1 : Different alternative for yield/acreage in relation to rainfall

to be a more reasonable agronomic assumption than the symmetric impact of the normal curve.

Taking logarithm on both sides, we get:

$$\log\left(\frac{O_1}{A_1}\right) = \text{Constant} - cR_g + p \log R_g \qquad \text{....(8)}$$

It is easy to derive the optimum growing season rainfall from the estimated coefficients of (8)—yields is at a maximum when R_g equals $R^*_g = p/c$. On overall agronomic considerations, too, the gamma formulation has more attractive properties. We consequently selected the gamma type specification for our analysis.

ACREAGE RESPONSE

Using the adjustment lagged model as the basic frame of analysis, the acreage response relationship in the present study was estimated with the help of the following equation:

$$\begin{aligned}\log PAC_t = \; & a_0 + b_1 NIR_{t-1} + b_2 IOR_{t-1} + b_3 OPR_{t-1} \\ & + b_4 R_s + b_5 \log R_s + b_6 PAC_{t-1}\end{aligned}$$

The specification of the various variables used for estimating the supply response of the acreage under the selected crops are discussed as below:

PAC_t : Per cent area under the crop considered at period (t). It is measured as the gross area under the selected crop as a percentage of gross cropped area at period (t).

PAC_{t-1}: Per cent area under the crop considered at period (t–1).It is measured as the area under the selected crop as a percentage of gross cropped area at period (t–1).

NIR_{t-1} : Irrigation ratio at period (t–1). It is measured as the net sown area irrigated as a per cent of net sown area at period (t–1).

IOR_{t-1} : Input-output ratio at period (t–1). It is measured as the input price (N-fertilizer) to output price of the selected crop at (t–1) period.

OPR_{t-1}: Output-price ratio at period (t–1). It is measured as the output price of the selected crops at (t–1) period. In case of more than one competing crops, simple average of the competing crops were taken.

The selected competing crops are as follow:

Main crop	*Competing crop (s)*
Wheat	Toria, Sugarcane
Paddy	Maize, Sugarcane
Cotton desi	Cotton american, Paddy
Cotton american	Sugarcane, Paddy
Sugarcane	Wheat, Paddy
Rapeseed and mustard	Wheat

R_s : Rainfalls during the pre sowing months of the selected crops. The selected months for the crops considered are as follow:

Crop(s)	*Month(s)*
Wheat	October
Paddy	May
Cotton	May
Sugarcane	March
Rapeseed & mustard	September

RESULTS AND DISCUSSIONS

The resulting estimates of the acreage response of the selected crops in the districts under considerations are reported in Table 6.1. Zero order correlation matrix (Annex Table 6.4 through 6.8) reveals that there was a high degree of correlation among some of the explanatory variables, that is, there occurred the problem of multicollinearity. Therefore, it was decided to computerise the regression analysis for different possible combinations (considering part of the variable) in order to find out the significant variables and the most appropriate regression model. The explanatory power of the fitted regression models was found to be highly satisfactory because the

Table 6.1: Structural coeficients of area response of selected crops in Punjab

Crop(s) Code	Intercept	Estimated value of the regression coefficient on						$\overline{R}^2$	Durbin-Watson Statistic
		NIR_{t-1}	OPR_{t-1}	IOR_{t-1}	R_s	log R_s	PAC_{t-1}		
WHL	0.585	0.068 (0.521)	-0.103‡ (3.643)	x	0.008* (2.181)	–0.012 (1.671)	0.581‡ (3.071)	0.5049	1.5341
WLA	–0.421	1.049‡ (4.075)	x	–0.049* (1.847)	0.015‡ (5.149)	x	x	0.7306	2.2165
PHL	–39.359	20.430‡ (5.743)	–0.139 (1.457)	x	0.036 (1.173)	x	x	0.7269	1.4589
CDF	0.407	–0.225 (0.354)	0.086 (0.314)	x	–0.065 (2.013)	x	0.844‡ (7.719)	0.8222	1.8544
CDB	2.527	–1.193 (1.489)	0.058 (1.499)	x	0.026 (0.921)	x	0.638‡ (4.599)	0.7552	1.7039
CAF	2.147	–0.678 (3.635)	0.065 (0.907)	x	–0.018‡ (4.186)	0.077‡ (4.387)	0.223 (1.139)	0.7031	2.4993
CAB	2.745	–1.336‡ (2.792)	x	–0.011 (0.851)	–0.204‡ (3.299)	–0.095 (0.514)	0.832‡ (10.877)	0.9052	2.7788
SGP	2.126	–1.087† (2.212)	x	x	0.002 (0.053)	x	0.508‡ (3.464)	0.6465	1.8824
RSA	15.654	–7.461‡ (3.458)	0.706‡ (3.967)	–0.861‡ (5.472)	x	x	x	0.6679	1.5054

For description of the crop(s) code,.please refer to the text.
Figures in parentheses are the 't' value of the respective parameters.
‡ implies significance at 0.01 level.
† implies significance at 0.05 level.
* implies significance at 0.10 level.

value of coefficient of determination corrected for degree of freedom is more than 0.70 in most of the cases. The computed Durbin-Watson statistic indicates the absence of auto correlation in most of the cases.

Irrigation Intensity

Under the assumption of partial adjustment and static expectation, irrigation intensity turns out to be the most important stimulus potent affecting area under wheat in Amritsar with a unitary elasticity and area under paddy in Ludhiana. Irrigation, however, turns out to be a non significant variable affecting area under wheat in Ludhiana. On the other hand, irrigation intensity adversely affected the area under cotton desi, cotton american, sugarcane and rapeseed and mustard in the selected districts (significantly for latter three crops) of the study area. The reason is quite obvious. As the irrigation infrastructure improves, farmers will shift to the other crops, namely, wheat and paddy which are less risky and have its mature market.

Rainfall

Rainfall have a monotonic effect on the proportion of wheat acreage in Amritsar and paddy acreage in Ludhiana. This is not surprising. Amongst all the crops grown in Amritsar, no crops other than wheat, can bear the damage of excess rains. At the most, this will only delay the sowing of wheat. Hence if there is excessive rains, farmers would switch out of the other crops into wheat as the best alternative. Likewise, in the rainy season (*kharif*) amongst all the crops grown in the rainy season in Ludhiana, paddy has the highest water requirements. Hence if there is excessive rains, farmers would switch out of other crops into paddy as the best alternative.

However, for area response of wheat for Ludhiana, non-monotonic type of specification—gamma type—of rainfall gave the better results. Further, rainfall has a monotonic effect on the area under cotton desi in Ferozepur and Bathinda, though adverse in case of former. Likewise, rainfall has a monotonically beneficial effect on area under sugarcane in Patiala, though it was not significant in statistical sense. In case of cotton american, gamma type specification of rainfall gave the better results and showed adverse effect. Thus, as the chances of rainfall improves, farmers will switch over to the competing crop, namely, paddy as the best alternative.

Delayed Adjustment

The coefficient of lagged acreage was found to be highly significant for all the crops under consideration of the selected districts in the study area. Its large value, generally indicates farmers slow adjustment. It takes almost ten to twenty years for 95 per cent of the effect of the price changes to be realised in the study area. Its main cause, as is evident from the general psychology of the farmers, is that they follow the tradition in determining the appropriate cropping pattern. They do not easily decide to reduce the area under the crop in comparison with the area they sowed in the preceding year. In this connection, their own requirements of food, ignorance about determining the appropriate cropping pattern owing to illitercy and other traditional and social standards adopted by the farmers and the lack of suitable competing crops, are the possible causes. Thus, various techno-institutional and subjective factors seem to have a great deal of influence on the farmers decision making acreage allocation

Price Ratios

Nevertheless, the farmers are clearly responsive to changes in their relative input output prices. The price response or acreage under paddy suggests that the doubling of the price of paddy, *ceteris paribus*, will lead to 13.9 per cent rise in acreage in Ludhiana. Likewise, the doubling of wheat price, *ceteris paribus*, will lead to 4.9 per cent and 10.3 per cent increase in area in Amritsar and Ludhiana respectively.

On the other hand, the price response of the market crops, considered in the study, namely, cotton desi, cotton american, sugarcane and rapeseed and mustard with respect to the output prices, *ceteris paribus*, will lead to a decline in its acreage.

YIELD RESPONSE

The yield response in this present study was estimated with the help of following equation:

$$\log Y_t = a_0 + b_1\ GIR_{t-1} + b_3\ IOR_{t-1} + b_3 R_g + b_4 \log R_g$$

The specification of the various variables used for estimating the supply response of the yield of different crops for the present study are discussed below:

Y_t : Per hectare yield (in kg) of the selected crop(s) during period (t).

GIR_{t-1} : Irrigation ratio at period (t–1). It is measured as gross area irrigated as a per cent of gross cropped area during period (t–1).

IOR_{t-1} : Input-output ratio at period (t–1). It is measured as the ratio of input price (N-fertilizer) at period (t) to out put price of the selected crop(s) at period (t–1).

R_g : Rainfall during the growing season of the selected crops. The selected months for the crops considered are as follow:

Crop(s)	*Month(s)*
Wheat	December
Paddy	July and August
Cotton	July
Sugarcane	April and May
Rapeseed & mustard	

The resulting estimates of the yield response of the selected crops of the districts under considerations are reported in Table 6.2. A comparison of the two types of specification of rainfall, namely, monotonic and gamma types, indicates that by specifying rainfall in a careful fashion, the value of $(\bar{R})^2$ and t- values have improved significantly.

The resulting estimates reported in Table 6.2 reveals that with the exception of wheat of Ludhiana and rapeseed and mustard of Amritsar, the effect of rainfall was not monotonic. The rainfall optimum was computed to be 9.4 cm for Paddy of Ludhiana; 1.52 cm for Wheat of Amritsar; 81 cm for Cotton desi of Ferozepur; 19 cm for Cotton desi of Bathinda; 35 cm

Table 6.2: Structural coefficients of yield response of selected crops in Punjab

Crop(s) Code	Intercept	Estimated value of the regression coefficient on				$\overline{R}^2$	Durbin-Watson Statistic
		GIR_{t-1}	IOR_{t-1}	R_g	log R_g		
PHL	–1.814	2.797‡ (3.901)	0.286 (1.653)	0.003* (1.867)	–0.028† (2.572)	0.5892	1.3707
WHL	0.367	1.654‡ (5.191)	–0.195 (1.771)	–0.009 (0.769)	x	0.6731	1.9318
WLA	–11.223	7.405‡ (5.207)	–0.262 (1.397)	0..052‡ (4.188)	–0.079† (2.605)	0.4054	1.8682
CDF	7.031	–2.473‡ (2.582)	0.065 (0.216)	–0.003 (0.876)	0.241 (1.507)	0.3684	2.4621
CDB	6.612	–2.115† (2.186)	0.025 (0.104)	0.011† (2.243)	–0.207* (1.879)	0.3206	1.7835
CAF	6.489	–1.953† (2.221)	0.037 (0.138)	0.006* (2.025)	–0.215 (1.461)	0.3383	2.6067
SGP	1.671	1.131‡ (3.221)	–0.274‡ (2.579)	–0.112 (0.786)	0.051* (2.014)	0.6304	1.1635
RSA	0.389	1.249 (0.297)	–0.172 (0.511)	0.011 (1.725)	x	0.1799	1.0911

For description of the crop(s) code, please refer to the text.
Figures in parentheses are the 't' value of the respective parameters.
‡ implies significance at 0.01 level.
† implies significance at 0.05 level.
* implies significance at 0.10 level.

for Cotton american of Ferozpur and 28 cm for Sugarcane of Patiala of the growing months of the respected crop(s) under considerations. In case of wheat crop of Ludhiana and rapeseed and mustard of Amritsar, rainfall has a monotonic effect on its yield, though not significant in both the cases. The adverse effect in case of wheat crop of Ludhiana may be attributed to the low water requirement of the crop and the availability of strong irrigation net work. The beneficial effect in case of rapeseed and mustard of Amritsar, though not significant, may be attributed to the fact that the crop is grown mainly under rainfed conditions and that too on the marginal land, in addition to the low water requirement of the crop.

The elasticity of yield with respect to prices of fertilizer (in terms of N) or the lagged price of the out-put turns out to be higher than the acreage corresponding elasticities. These elasticities were computed to be 0.286 for paddy of Ludhiana; –0.195 and –0.262 respectively for wheat of Ludhiana and Amritsar; 0.065 and 0.025 for cotton desi respectively for Ferozepur and Bathinda; –0.274 for sugarcane in Patiala and –0.172 for rapeseed and mustard in Amritsar.

Finally, irrigation turns out to be a much more potent stimulus. The impact of one per cent increase in gross area irrigated lead to 2.8 per cent increase in the yield of paddy. Likewise, one per cent increase of gross irrigated area had lead to an increase of 1.7 and 7.4 per cent increase in the yield of wheat in Ludhiana and Amritsar respectively; 1.13 per cent increase in the yield of sugarcane in Patiala and 1.25 per cent increase in the yield of rapeseed and mustard in Amritsar. Likewise, the irrigation ratio turns out to be a much more important variable affecting the yield of cotton adversely (of both, namely, cotton desi and cotton American) in both the districts under consideration, viz., Ferozepur and Bathinda in the study area.

OUTPUT RESPONSE

The output response of the selected crops has been derived as a reduced form because it is the output response which is more important from the planning point of view. Incorporation of rainfall in a monotonic as well as non-monotonic fashion in the supply functions gave the output elasticities as reported in Table 6.3. Nevertheless, the prices are obviously an effective policy instrument in influencing output. Farmers, are, clearly, responsive to changes in their relative input-output prices.

POLICY IMPLICATIONS

The results obtained in the present study lead to important implications that seem to be relevant from the point of view of policy formulations. The estimated price elasticities turn out to be fairly low with respect to acreage change but significantly high price elasticities emerge for yield as well as for output. Switching out of wheat and paddy into an alternative crop is not easy perhaps due to the agronomic characteristics. Farmers are, however, clearly, responsive to the change in their relative input output prices. An ideal price policy has to consider, therefore, change in relative prices rather than absolute change. Government can bring about suitable changes in the cropping pattern by influencing changes in the relative profitability of the crop through regulation of relative prices. A well designed agricultural price policy can be used to induce the farmers to plan the production of different crops in line with the estimated requirements in the country.

The study cautions against making ad hoc decisions about the regulation of prices of individual crops and leaving the competing or the substitute crops uncontrolled. The government has a two edged weapon at its disposal to influence change in the cropping pattern conducive to the overall needs of the country. First by regulating the relative prices of the crops, and, second by regulating with suitable taxes and subsidies for the inputs used of a particular crop.

The study has brought out the fact that the area under the crop in the preceding year had a significant influence on the area under the crop in a year. Its main cause, as is evident from the general psychology of the farmers, is that they follow the tradition in determining the appropriate cropping pattern. In order to bring effective adjustment in acreage

Table 6.3: Output elasticities of selected crops

Particular	Estimated value for							
	PHL	WHL	WLA	CDF	CDB	CAF	SGP	RSA
Elasticity w.r.t.								
1. Output price	0.4256	0.2055	0.3109	–0.1514	–0.0834	–0.1024	0.2743	0.8613
2. Input price	0.2864	–0.1952	–0.2571	0.0652	0.0251	0.0372	–0.2743	–1.0332
3. Substitute crop price	0.1392	0.1031	x	0.0862	0.0583	0.0652	x	0.7061

Output elasticities were derived as a reduced form.
For description of the crop code(s), please see text.

allocation, the support prices for various crops must be announced well before the sowing season and the prices thus announced, should carry a long run guarantee. This policy will not only enable the farmers to plan their production programmes better but also help to correct the inter commodity imbalance to some extent.

The varietal improvements in the selected crops(s) and the adoption of the varieties by the farmers having comparative advantages, technological, institutional and others, will be important for maintaining production at the desired level. Breeders have to shoulder this responsibility with utmost care. Besides, the government needs to take some other steps also. Education and training of the farmers, crop protection and crop insuracne schemes, regulation of the market, creation of co-operatives, and warehousing facilities can be some of the steps in this direction.

Finally, an integrated production approach covering the supply of adequate inputs at reasonable prices, well distributed supply of water; evolving improved varieties of seeds which may not be susceptible to the vagaries of monsoons, and a judicious price policy linking input price with the output prices is urgently called for. Infact, price policies must go hand in hand with the policies relating to the production technology.

REFERENCES

1. Land will be a reasonably good proxy for production so long as it is a major input and other inputs are closely complementary to it. In a situation, where it is not so and use of non-land inputs should be explicitly considered while determining supply response. Moreover, agro climatic conditions may restrict allocation of land among different crops but the allocation of other inputs like labour, fertilizer, pesticides, irrigation etc. may not be subject to a similar restriction.
2. Some of these studies are Askari and Cummings (1976); Krishna and Raychaudhuri (1980); Parikh and Narayana (1981); Lahiri and Roy (1984); Grewal and Rangi (1976); Sidhu (1979) and Kainth and Kang (1984).
3. Professor Nurkse has suggested that the demonstration effect on the international plane has an important influence on the consumption in developing areas. Similarly, the improved consumption pattern of the urban sector will influence the consumption pattern of the rural sector.

Annex Table 6.1
Estimated value of the regression coefficients

Model:- $\text{Log } A_t = \alpha + \beta \text{ Log } A_{t-1}$

Crop(s) code	Pure Intercept	Estimated value of regression coefficient on $A_{(t-1)}$	$\bar{R}^2$	D.W.
WHL	1.0021	0.3984 (1.5545)	0.0902	1.6854
WLA	0.6760	0.5847† (2.8593)	0.3373	2.2231
PHL	0.1466	0.9399‡ (16.9312)	0.9533	2.6291
PLA	0.1497	0.9036‡ (9.8527)	0.8728	2.2607
CAB	0.2212	0.8426‡ (7.7148)	0.8069	2.3485
CAF	0.1833	0.8162† (2.7527)	0.3196	2.0667
CDB	0.0287	0.9175‡ (6.5297)	0.7484	2.1021
CDF	–0.0204	0.8757‡ (7.5552)	0.8003	2.1817
SGP	0.0299	0.7553‡ (5.1506)	0.6458	1.7519
RSA	0.1912	0.6681‡ (3.2298)	0.4025	1.9649

For description of crop(s) code, please see text
‡ implies significance at 0.01 level
† implies significance at 0.05 level
* implies significance at 0.10 level

Annex Table 6.2
Estimated value of the regression coefficients

Model:- $\text{Log } P_t = \alpha + \beta \text{ Log } P_{t-1}$

Crop(s) code	Pure Intercept	Estimated value of regression coefficient on $P_{(t-1)}$	$\bar{R}^2$	D.W.
Wheat	0.2159	0.1585 (0.5471)	–0.0527	1.7618
Paddy	0.1063	0.2936 (1.1166)	0.0173	1.3762
Cotton american	–0.0034	0.2917 (1.1411)	0.0211	1.5211
Cotton desi	–0.0735	0.5557 (1.8232)	0.1424	1.8352
Sugarcane	–0.1429	0.1745 (0.6491)	–0.0431	1.3942
Rapeseed & mustard	–0.1879	0.3477 (1.2699)	0.0419	1.4829

‡ implies significance at 0.01 level
† implies significance at 0.05 level
* implies significance at 0.10 level

Annex Table 6.3
Estimated value of the regression coefficients

Model:- $\text{Log } Y_t = \alpha + \beta \log Y_{t-1}$

Crop(s) code	Pure Intercept	Estimated value of regression coefficient on $Y_{(t-1)}$	$(\bar{R})^2$	D.W.
WHL	1.1359	0.6780† (2.3581)	0.2463	1.7227
WLA	1.1145	0.6749† (2.5872)	0.3152	1.7011
PHL	0.8886	0.7520‡(11.4827)	0.9033	2.6107
PLA	1.9750	0.4112 (1.7329)	0.1252	1.8324
CAB	2.1666	0.1584 (0.5593)	0.0514	1.8717
CAF	2.2088	0.1359 (0.4792)	0.0581	1.9151
CDB	1.6840	0.3246 (1.2329)	0.0358	1.7976
CDF	1.7624	0.2830 (1.0472)	0.0068	1.8822
SGP	2.1678	0.4202* (1.7782)	0.1336	1.6907
RSA	0.9420	0.6758† (2.7875)	0.3259	1.9005

For description of crop(s) code, please see text
‡ implies significance at 0.01 level
† implies significance at 0.05 level
* implies significance at 0.10 level

Annex Table 6.4
Zero order correlations matrix of selected variables affecting area under wheat

	NAI	OPR	R_s	$\log R_s$	IOR	GCA_{t-1}
			Ludhiana			
OPR	0.4846					
R_s	0.3922	0.4278				
$\log R_s$	0.0373	0.5182	0.5902			
IOR	0.3807	0.2691	–0.0458	–0.0074		
GCA_{t-1}	–0.2686	0.2057	–0.0014	0.0318	–0.5289	
GCA_t	–0.2357	–0.4875	–0.0186	–0.4727	–0.1084	0.3982
			Amritsar			
OPR	0.6195					
R_s	0.2934	–0.1152				
$\log R_s$	0.2259	–0.0801	0.7455			
IOR	0.4161	0.3101	0.4429	0.2829		
GCA_{t-1}	0.8099	0.5607	0.3043	0.0836	0.2735	
GCA_t	0.6809	0.2050	0.6621	0.2163	0.4024	0.6273

For description of the variables, please see text

Annex Table 6.5
Zero order correlations matrix of selected variables affecting area under paddy

	NAI	OPR	R_s	log R_s	IOR	GCA_{t-1}
			Ludhiana			
OPR	0.1624					
R_s	−0.2234	−0.1641				
log R_s	−0.0629	−0.3041	0.7514			
IOR	0.1814	0.6480	0.1214	−0.2148		
GCA_{t-1}	0.9506	0.2058	−0.1729	−0.0206	0.1674	
GCA_t	0.9258	0.1904	−0.1502	0.0335	0.1771	0.9854
			Amritsar			
OPR	0.4603					
R_s	−0.0521	−0.0488				
log R_s	−0.1222	−0.1207	0.9410			
IOR	0.3242	0.6481	0.2281	0.2198		
GCA_{t-1}	0.9727	0.3891	−0.0201	−0.0669	0.2913	
GCA_t	0.9618	0.3239	−0.0377	−0.8960	0.3895	0.9484

For description of the variables, please see text

Annex Table 6.6
Zero order correlations matrix of selected variables affecting area under cotton american

	NAI	OPR	R_s	log R_s	IOR	GCA_{t-1}
			Bathinda			
OPR	0.0875					
R_s	−0.2179	0.0066				
log R_s	−0.3421	−0.0928	0.8764			
IOR	0.0179	0.3919	0.1885	0.0334		
GCA_{t-1}	0.6574	−0.1695	−0.4448	−0.5194	0.2414	
GCA_t	0.4969	−0.1491	−0.5682	−0.6632	0.1789	0.9136
			Ferozepur			
OPR	−0.0765					
R_s	−0.0495	0.1829				
log R_s	−0.3609	0.0753	0.8196			
IOR	0.1576	0.3919	0.0154	0.1234		
GCA_{t-1}	−0.7739	−0.1324	0.2275	0.3343	−0.1367	
GCA_t	−0.8643	0.1288	−0.0259	0.3578	−0.0019	0.6157

For description of the variables, please see text

Annex Table 6.7
Zero order correlations matrix of selected variables affecting area under cotton desi

	NAI	OPR	R_s	log R_s	IOR	GCA_{t-1}
			Bathinda			
OPR	–0.3423					
R_s	–0.2179	0.3459				
log R_s	–0.3420	0.4316	0.8764			
IOR	–0.2388	0.4881	0.3721	0.2755		
GCA_{t-1}	–0.4644	0.8078	0.2997	0.4226	0.2887	
GCA_t	–0.4385	0.8311	0.4423	0.5564	0.1625	0.8591
			Ferozepur			
OPR	–0.5303					
R_s	–0.0495	0.3070				
log R_s	–0.3601	0.3524	0.8196			
IOR	–0.3126	0.5313	0.1431	0.3289		
GCA_{t-1}	–0.6082	0.7685	0.4255	0.4979	0.2368	
GCA_t	–0.5855	0.6772	0.2055	0.4159	0.2026	0.8979

For description of the variables, please see text

Annex Table 6.8
Zero order correlations matrix of selected variables affecting area under sugarcane and rapeseed and mustard

	NAI	OPR	R_s	log R_s	IOR	GCA_{t-1}
			Sugarcane (Patiala)			
OPR	–0.0705					
R_s	0.2799	–0.3361				
log R_s	–0.0269	–0.2668	0.7638			
IOR	0.1913	0.7972	–0.1645	–0.1211		
GCA_{t-1}	–0.7480	0.5768	–0.3073	–0.1006	0.3791	
GCA_t	–0.7521	0.1192	–0.2388	–0.0625	0.0605	0.8202
			Rapeseed and Mustard (Amritsar)			
OPR	–0.4767					
R_s	–0.1577	–0.0817				
log R_s	0.0445	–0.3364	0.8346			
IOR	–0.2761	0.6545	0.2315	0.1868		
GCA_{t-1}	–0.6949	0.6851	0.0581	–0.1909	0.3196	
GCA_t	–0.6476	0.2986	–0.0071	–0.2486	–0.2961	0.7125

For description of the variables, please see text

Annex Table 6.9

Zero order correlation matrix of factors affecting yield of wheat

	GIR	IOR	R_g	log R_g	Y_{t-1}
Ludhiana					
IOR	0.5681				
R_g	0.1430	0.4847			
log R_g	0.2648	0.1484	0.7605		
Y_{t-1}	0.5854	0.0406	0.1034	0.0047	
Y_t	0.5783	–0.0489	–0.3038	–0.4889	0.5468
Amritsar					
IOR	0.2554				
R_g	–0.7162	0.0231			
log R_g	–0.4462	0.1493	0.8801		
Y_{t-1}	0.2269	–0.0081	0.2161	0.1541	
Y_t	0.6652	–0.0800	–0.4153	–0.3229	0.6109

For description of variables, please see text

Annex Table 6.10

Zero order correlation matrix of factors affecting yield of paddy

	GIR	IOR	R_g	log R_g	Y_{t-1}
Ludhiana					
IOR	0.3983				
R_g	0.1404	–0.1652			
log R_g	0.0833	–0.1198	0.9874		
Y_{t-1}	0.8657	0.5966	–0.0075	–0.0494	
Y_t	0.7893	0.5492	–0.0682	–0.1032	0.9542
Amritsar					
IOR	0.5072				
R_g	0.1735	0.1447			
log R_g	0.2395	0.2104	0.8091		
Y_{t-1}	0.5082	0.3053	0.0148	0.0954	
Y_t	0.2336	–0.1098	0.0203	0.0332	0.4335

For description of variables, please see text

Annex Table 6.11
Zero order correlation matrix of factors affecting yield of cotton american

	GIR	IOR	R_g	log R_g	Y_{t-1}
		Bathinda			
IOR	0.4540				
R_g	-0.1009	0.2562			
log R_g	0.0732	0.2481	-0.3556		
Y_{t-1}	-0.5536	-0.1134	0.1958	-0.0869	
Y_t	-0.6111	-0.2097	0.1450	-0.0552	0.1534
		Ferozepur			
IOR	0.3752				
R_g	0.3170	0.2391			
log R_g	0.3938	0.1221	0.9528		
Y_{t-1}	-0.6818	0.0485	-0.1723	-0.3001	
Y_t	-0.5407	-0.1002	-0.0096	-0.0999	0.1317

For description of variables, please see text

Annex Table 6.12
Zero order correlation matrix of factors affecting yield of cotton desi

	GIR	IOR	R_g	log R_g	Y_{t-1}
		Bhatinda			
IOR	-0.2048				
R_g	0.1106	0.4442			
log R_g	0.0109	0.2829	0.9290		
Y_{t-1}	-0.5005	0.2989	0.0178	-0.0382	
Y_t	-0.5159	0.2647	0.0786	0.0547	0.3236
		Ferozepur			
IOR	-0.2864				
R_g	0.3169	0.0192			
log R_g	0.3938	-0.1406	0.9528		
Y_{t-1}	-0.8361	0.2797	-0.1958	-0.3033	
Y_t	-0.5852	0.1826	-0.0639	-0.0942	0.2789

For description of variables, please see text

Annex Table 6.13

Zero order correlation matrix of factors affecting yield of sugarcane in Patiala

	GIR	IOR	R_g	log R_g	Y_{t-1}
IOR	0.1882				
R_g	0.3772	–0.0142			
log R_g	0.3441	0.1393	0.8051		
Y_{t-1}	0.5071	0.3861	0.0491	–0.0752	
Y_t	0.6020	–0.3231	0.4012	0.4095	0.4424

For description of variables, please see text

Annex Table 6.14

Zero order correlation matrix of factors affecting yield of rape-seed & mustard in Amritsar

	GIR	IOR	R_g	log R_g	Y_{t-1}
IOR	–0.1540				
R_g	0.3337	–0.4152			
log R_g	0.4431	–0.2470	0.8398		
Y_{t-1}	0.3678	0.0988	0.3479	0.1857	
Y_t	0.2056	–0.2801	0.3931	0.3012	0.6116

For description of variables, please see text

BIBLIOGRAPHY

Abraham, T.P., "Isolation of effects of weather on productivity including other risks such as damages by pests and diseases", *Journal of Indian Society of Agricultural Statistics*, Vol. 17, No. 2, 1965.

Abraham, T.P. and S.P. Raheja, "An analysis of growth of production of rice and wheat crops in India", *Indian Journal of Agricultural Economics*, Vol. 22, No. 3, July–September 1967.

Acharya, S.S. and Satish Bhatia, "Acreage response to price, yield and rainfall changes in Rajasthan", *Agricultural Situation in India*, Vol. 29, No. 4, April 1974.

Agarwal, Ranjana, R.C. Jain and M.P. Jha, *Effects of Weather on Rice Yield and Its Forecast—Puri District*, Division of Crop Forecasting Methodology, Indian Agricultural Statistics Research Institute, New Delhi, 1984.

Agarwal, Ranjana, R.C. Jain, M.P. Jha and D. Singh, "Forecasting of rice yield using climatic variables", *Indian Journal of Agricultural Science*, Vol. 50, No. 9, 1980.

Alsberg, C.L. and E.F. Griffing, "Forecasting wheat yield from the weather elements of an unsolved problem", *Wheat Studies of Food Research Institute*, Vol. 5, No. 1, November 1928.

Amemiya, T., "Selection of regressors", *International Economic Review*, 1980.

Askari, H. and J.T. Cummings, *Agricultural Supply Response: A Survey of the Econometric Evidence*, Praeger, New York, 1976.

Ayyar, P.S. Harihara, G. Rama Rao and J.R. Banerji, "Influence of day degree on the elongation of sugarcane", *Indian Sugarcane Journal*, Vol. 19, No. 4, July–September 1965.

Bal, H.K. and D. Raghava Rao, "Long run and short run projections of area under wheat in the Punjab state", *Journal of Indian Statistical Association*, Vol. 25, No. 2, 1973.

Bapna, S.L., *Aggregate Supply Response of Crops in a Developing Region*, Sultan Chand and Sons, New Delhi, 1981.

Bardhan, K., "Relative prices and allocation of land and other inputs among competing crops", in A.M. Khusro (Ed.) *Readings in Agricultural Development*, Allied Publishers Private Ltd., 1968.

Bardhan, P.K., "Size productivity and return to scale: an analysis of farm level data in Indian agriculture", *Journal of Political Economy*, Vol. 81, No. 6, November-December 1973.

Barker, Randolf and F.S. Bernard, "Estimation and aggregation of firm supply functions", *Journal of Farm Economics*, Vol. 47, August 1965.

Batra, M.M., *Agricultural Production Prices and Technology*, Allied Publishers, Bombay, 1978.

Batra, M.M., "Supply Response of a subsistence crop under traditional and new technologies", *Indian Economic Review*, Vol. 11, No. 2, New Series, October 1976.

Baur, P.T. and B.S. Yamey, "A Case study of response to price in an underdeveloped country", *The Economic Journal*, Vol. 69, No. 276, December 1959.

Bean, Louis H., "A new approach to statistical forecasting of next year's weather and crops", *Agricultural Situation in India*, Vol. 23, No. 12, March 1969.

Bean, Louis H., "Conceptualising the supply relation in agriculture", *Journal of Farm Economics*, Vol. 37, December 1955.

Bean, Louis H., "Symmetry in weather and crop statistics", *Journal of the Indian Society of Agricultural Statistics*, Vol. 22, No. 2, December 1970.

Bedekar, V.C., G. Appa Rao and M.L. Madnani, *Forecasting Wheat Yield in Himachal Pradesh with the Help of Weather Elements*, Meteorological Monograph (mimeo.), 1973.

Bedekar, V.C:, G. Appa Rao and G. Ramachandran, *Forecasting Wheat Yield in East Rajasthan with the help of Weather Parameters*. Meteorological Monograph (mimeo.) 1973.

Behrman, J.R., *Supply Response in Underdeveloped Agriculture: Contributions to Economic Analysis*, North Holland Publishing Company, Amsterdam, 1968.

Behrman, J.R., *Adoption of new products and of new factors in response to market incentives in peasant agriculture: an economic investigation of Thai corn and Kanaf supply responses in the post war period*, Discussion Paper No. 45, The Wharton School of Finance and Commerce, Department of Economics, University of Pennsylvania, Philadelphia, 1967.

Bhagwati, J.N. and S. Chakravarthy, "Contributions to Indian economic analysis—a survey", *The American Economic Review*, Vol. 59, No. 4, Part 2, September 1969 (Supplement).

Bhakthavathsalu, C.M. and C. Balsubramanian, "A note on the statistical assessment of probability of rainfall and its agricultural significance,", *Madras Agricultural Journal*, Vol. 45, No. 3, 1959.

Bhargava, P.N. et al., "A study of the occurrence of rainfall in Raipur district with the help of Markov chain model", *Indian Journal of Agricultural Economics*, Vol. 25, No. 2, March-April 1973.

Bhargava, P.N. et al., "A study of the recurrence of rainfall deficiency in relation to rice crop", *JNKVV Journal of Research*, Vol. 8, Nos. 3 and 4, 1974.

Bhargava, P.N. et al., *Statistical Studies on the Behaviour of Rainfall in a Region in Relation to a Crop*, Institute of Agricultural Research Statistics, ICAR, New Delhi, 1977.

Bhargava, P.N. et al., "Influence of rainfall on crop production", *JNKVV Research Journal*, 1974.

Bhattacharya, N. and T. Maitra, *An Analysis of Engel Curves Based on NSS Households Budget Data From 7th to 22nd Rounds*, Technical Report No. Econ/9/70, Indian Statistical Institute, Calcutta, 1970.

Blyn, George, *Agricultural Trends in India: 1891–1947: Output Availability and Productivity*, University of Pennsylvania, Pennsylvania, 1966.

Blyn, George, "Measurement of growth rates in agriculture", *Indian Journal of Agricultural Economics*, Vol. 22, No.1, January–March 1967.

Box, G.E.P. and D.R. Cox, "An analysis of transformation", *Journal of Royal Statistical Society (Series B)*, Vol. 26, 1964.

Breimyer, Harold F., "Nature's felicity in supply response analysis: comment", *American Journal of Agricultural Economics*, Vol. 52, February 1970.

Brewster, J. and M. Parson, "Can prices allocate resources in agriculture?", *Journal of Farm Economics*, Vol. 28, November 1946.

Brown, R.L., et al., "Techniques for testing the constancy of regression relationship over time", *Journal of Royal Statistical Society*, 1975.

Brownke, O.H. and W. Gainer, "Farmer's price anticipations and planning", *Journal of Farm Economics*, Vol. 31, 1949.

Burt, Oscar R., "Positivistic measures of aggregate supply elasticities: some new approaches: comment", *American Journal of Agricultural Economics*, Vol. 53, 1971.

Burt, Oscar, R., "Positivistic measures of aggregate supply elasticities: some new approaches—rejoinder "*American Journal of Agricultural Economics*, Vol. 54, August 1972.

Candler, Wilfred, "An aggregate supply functions for New Zealand wheat", *Journal of Farm Economics*, Vol. 39, 1957.

Cassels, John, "The nature of statistical supply curves", *Journal of Farm Economics*, April 1933.

Chan, F., "A preliminary study of supply response of estates between 1948 and 1959", *Malayan Economic Review*, Vol. 7, No. 2, October 1962.

Chand, Mahesh, "Agricultural terms of trade and economic growth", *Indian Journal of Agricultural Economics*, Vol. 13, No. 1, 1958.

Chaudhri, S.C., "Producers price policy", *Agricultural Situation in India*, Vol. 17, No. 5, Annual Number, August 1962.

Clark, R., "The economic determinants of jute production", *Monthly Bulletin of Agricultural Economics and Statistics (FAO)*, Vol. 6, No. 9, September 1957.

Colin, Clark, *The Economics of Irrigation*, Pergamon Press, Oxford, 1967.

Colin, Clark and Haswell Margaret, *The Economics of Subsistence Agriculture*, Mac Millan and Co. Ltd., London, 1970.

Cummings, J.T., "The supply responsiveness of Indian farmer's in the post Independence period: major cereals and cash crops", *Indian Journal of Agricultural Economics*, Vol. 30, No. 1, January–March 1975.

Cummings, Ralph W. (jr) and S.K. Ray, "Foodgrains production: relative contribution of weather and technology", *Economic and Political Weekly*, Vol. 4, No. 39, September 27, 1969.

Dandekar, V.M., "Agricultural price policy", *Economic and Political Weekly*, Vol. 3, No. 2, March 16, 1968.

Dandekar, V.M. "Introduction to seminar on data base and methodology for study of growth rates in agtriculture", *Indian Journal of Agricultural Economics*, Vol. 35, No. 2, April–June 1980.

Dandekar, V.M., "Minimum support prices for foodgrains: guidelines for a policy and a programme", *Artha Vijnana*, Vol. 7, No. 4, December 1965.

Dalton, G., "Traditional prodution in primitive African economies", *The Quarterly Journal of Economics*, Vol. 76, No. 3, August 1962.

Dantwala, M.L., "Incentive and disincentives in Indian agriculture", *Indian Journal of Agricultural Economics?* Vol. 22, No. 2, April–June 1967.

Das, Gupta Sipra, *Agriculture: Producer's Rationality and Technical Change,* Asia Publishing House, Bombay, 1970.

Das, J.C., *Forecasting Yield of Principal Crops in India on the Basis of Weather—Paddy/Rice*, Indian Meteorology Department, Scientific Report No. 120, February 1970.

Das, J.C., *Forecasting Rabi Jowar Yield in Bijapur District of Mysore State with Weather Parameters*, Mimeo., Meteorological Monograph, 1972.

Das, J.C., et al., *Forecasting the Yield of the Principal Crops in India on the Basis of Weather—Paddy/Rice—Gangetic West Bengal, Uttar Pradesh, Punjab and Haryana, Konkan, Madhya Pradesh East, Madhya Maharashtra,*

Orrisa and Vidarbha, Pre-published Report Nos. 138, 139, 161, and 170, Meteorological Monographs, Indian Meteorological Department, 1971.

Das, J.C. et al., *Forecasting Wheat in India with the Help of Weather Parameters—Haryana, Punjab, Uttar Pradesh, Bihar Plains and West Madhya Pradesh*, Pre-published Report Nos. 140 and 160, Indian Meteorological Department, 1971.

Das, J.C. et al., *Forecasting the Yield of Principal Crops in India on the Basis of Weather—Paddy / Rice, Mysore State*, Part II, Pre-published Report No. 124, Indian Meteorological Department, 1970.

Das, J.C. and G. Ramachandran, *Forecasting the Yield of Bajra on the Basis of Weather Parameters of Ahmedabad District*, Pre-published Report 141, Indian Meteorological Department, 1971.

Das, J.C. and P.P. Sajnani, "Locating critical periods when weather elements has significant influence on crop yields—applied to Telengana", paper presented at the *Seminar on Incidence of Aridity and Droughts in Andhra Pradesh*, Waltair, 1970.

Dastane, N.G., *Review of Work Done on Water Requirements of Crops in India*, Navabharat Prakashan, Poona, 1970.

Davidson, R. and J. Mackinnon, "Several tests for model specification in the pressence of alternative hypotheses", *Econometrica*, 1981.

Davis, F.E. and J.E. Pallesen, "Effects of the amount and distribution of rainfall and evaporation during the growing season on yield of corn and spring wheat", *Journal of Research*, Vol. 60, No. 1, January 1940.

Dayal, Ram, "Impact of climate on yield", *Agricultural Situation in India*, Vol. 14, No. 8, November 1959.

Dayal, Ram, "Impact of rainfall on crop yield and acreage", *Indian Journal of Agricultural Economics*, Vol. 20, No. 3, July–September 1965.

Dean, C.W. and E.O. Heady, "Changes in supply response and elasticity of hogs", *Journal of Farm Economics*, Vol. 40, No. 4, 1958.

Dean, E.R., "Economic analysis and African response to price", *Journal of Farm Economics*, Vol. 47, No. 2, May 1965.

Deasi, G.M. and Michael G.G. Schluter, "Generating employment in rural areas", *Seminar on Rural Development for the Weaker Sections*, Indian Society of Agricultural Economics, Bombay, 1972.

Devi, P. Kamla, "Response of acreage to change in price—a case study in Madras state", *The Economic Weekly*, Vol. 16, No. 38, 1964.

Devi, S. Uma, "Perennial crop supply response—the case of Indian natural rubber", *Indian Journal of Agricultural Economics*, Vol. 32, No. 4, 1977.

Dey, Amal Krishna, "Rates of growth of agriculture and industry," *Economic and Political Weekly*, Vol. 10, Nos. 25 & 26, June 21, 28, 1975.

Doll, John P., "An analytical technique for estimating weather indexes from meteorological measurements", *Journal of Farm Economics*, Vol. 49, No. 1, February 1957.

Dubey, J.P., "Weather indexes—a review of methods for constructing indexes of the effect of weather on crops", *Journal of the Indian Society of Agricultural Statistics*, Vol. 22, No. 2, December 1970.

Dutt, N.L., *Report on Survey of Sugarcane Research in India*, Indian Central Sugarcane Committee, New Delhi, 1965.

Durbin, J., "Testing for serial correlation to least square regression when some of the regressions are lagged dependent variables", *Econometrica*, Vol. 38, No. 3, 1970.

Durbin, J. and G.S. Watson, "Testing of serial correlation in least square regression", *Biometrica*, Vols. 37 and 38, Part I and II, 1950 and 1951.

Dutia, B.P., "Structure of relative prices of farm products in relation to their production and supply", Rapporteur's report and summary of group discussion, *Indian Journal of Agricultural Economics*, Vol. 20, No. 1, January–March 1965.

Eicher, C. and L. Witt (Eds.), *Agriculture in Economic Development*, McGraw Hill, New York, 1964.

Egbert Alvin, C., "The aggregate model of agriculture—empirical estimates and some policy implications", *Journal of Farm Economics*, Vol. 49, February 1967.

Falcon, W.P., "Farmer's responsiveness to price in subsistence economy—the case of West Pakistan", *Farm Economic Review*, Vol. 34, No. 3, 1964.

Feller, William, *An Introduction to Probability Theory and its Applications*, John Wiley and Sons, New York, 1966.

Fisher, F.M. and P. Temin, "Regional specialization in the supply of wheat in the United States: 1867–1914, "*Review of Economics and Statistics*, Vol. 52, No. 2, 1970.

Fisher, R.A., "Studies in crop variations–I: an examination of yield of dressed grain from broad balk", *Journal of Agricultural Science*, Vol. 11, No. 2, 1920.

Fisher, R.A., "The influence of rainfall on the yield of wheat at Rothamstead", *Philosophical Transactions of the Royal Society of London*, series B, No. 213, 1924.

Gangopadhyay, M. and R.P. Sarkar, "A correlation study on crop weather relationship—sugarcane", *Indian Journal of Sugarcane Research and Development*, Vol. 8, No. 1, October–December 1963.

Gangopadhyay, M. and R.P. Sarkar, "A curvilinear study of the effect of weather on the growth of sugarcane", *Indian Journal of Meteorology and Geophysiscs*, Vol. 15, No. 2, April 1964.

Gangopadhyay, M. and R.P. Sarkar, *Agricultural Meteorology,* mimeo., 1965.

Ganguli, B.N., "Agricultural development and price mechanism", in A.M. Khusro (Ed.) *Readings in Agricultural Development*, Allied Publishers, New Delhi, 1968.

Georgescu, Rogen N., "Economic theory and agrarian economies", *Oxford Economic Papers*, Vol. 12, February 1960.

Ghulam, Mohammad, "Some physical and economic determinants of cotton production in Pakistan", *Pakistan Development Review*, Vol. 3, Winter 1963.

Ghulam, Rabbani,, A.K.M., "Economic determinants of jute prodution in India and Pakistan", *Pakistan Development Review*, Vol. 2, No. 2, Summer 1965.

Gill, P.E., W. Murray and S.M. Picken, *The Implementation of Two Modified Newton Algorithms for Uncostrained Optimisation*, National Physical Laboratory Report, NAC, 24, UK.

Giri, A.K., "Impact of variations in rainfall and weather conditions on prodution of crops in West Bengal", paper presented at the *Symposium on Effects of Weather on Agri-*

cultural Prodution, University of Kalyani, Kalyani, West Bengal, February 1979 (mimeo).

Gislason, Conard, "Nature of aggregate supply function for agricultural products", *Journal of Farm Economics*, Vol. 34, February 1952.

Girilchies, Z., "Distributed lags: a survey", *Econometrica*, Vol. 35, 1965.

Girilchies, Z., "The estimates of the aggregate US farm supply function", *Journal of Farm Economics*, Vol. 42, May 1960.

Girilchies, Z., "Measuring inputs in agriculture: a critical survey, *Journal of Farm Economics*, Vol. 42, December 1960.

Grilichies, Z., "The sources of measured productivity growth", *Journal of Political Economy*, August 1963.

Grilichies, Z., "The demand for inputs in agriculture and derived supply elasticity", *Journal of Farm Economics*, Vol. 41, May 1959.

Grewal, S.S. and A.S. Kahlon, "A study of fluctuations in cane supply industry in Punjab", *Indian Sugar*, Vol. 16, No. 5, 1964.

Gupta, S.C. and A. Majid, *Producer's Response to Changes in Price and Marketing Policies: A Case Study of Sugarcane and Paddy in Uttar Pradesh*, Asia Publishing House, London, 1965.

Haller, J.G. and S. Wortman, "Expanding food production in hungry nations: the promise and problems", in Clifford M. Hardin (Ed.) *Overcoming World Hunger*, Prentice Hall, New Jersy, 1969.

Heady, Earl O., *Agricultural Policy Under Economic Development*, Iowa State University Press, Ames, 1962.

Heady, Earl O., "Uses and concept in supply analysis", in Earl O. Heady et al. (Eds.) *Agricultural Supply Functions: Estimating Techniques and Interpretation,* Iowa State University Press, Iowa, Ames, 1961.

Heady, Earl O. and J.L. Dillon, *Agricultural Production Function*, Ames, Iowa, 1961.

Heady Earl O. et al. (Eds.), *Economic Models and Quantitative Methods and Planning in Agriculture*, Iowa State University Press, Ames, 1966.

Hendricks, W.A. and J.C. Scholl, *Techniques in Measuring Joint Relationships: The Joint Effect of Temperature and Precipatation on Corn Yields*, The Agricultural Experiment Station of the North Carolina State College and US Department of Agricultural, Bureau of Agricultural Economics Coperation, N.C. State College Technical Bulletin 74, 1943.

Hendrix, W.E., "Factors associated with differences in levels and rates of change in agricultural output in underdeveloped countries", paper presented to the MIT conference on Agricultural Productivity, 1964 (mimeo.).

Henney, H.J., "Forecasting the yield of winter wheats even months prior to the harvest", *Journal of Farm Economics*, Vol. 14, No. 2, April 1932.

Herdt, Robert W., "The impact of rainfall and irrigation on crops in Punjab: 1907–1946", *Indian Journal of Agricultural Economics*, Vol. 27, No. 1, January–March 1972.

Hodges, J.A., "The effect of rainfall and temperature on corn yields in Kansas", *Journal of Farm Economics*, Vol. 13, No. 2, 1931.

Hooker, R.H., "Correlation of weather and crops", *Journal of the Royal Statistical Society*, No. 70, March 1907.

Hossein, A. and J.T. Cummings, "Supply response of farmers with heterogeneous land", *Indian Journal of Agricultural Economics*, Vol. 31, No. 1, January–March 1976.

Houseman, E.F., *Methods of Computing a Regression of Yield on Weather*, Iowa State College, Research Bulletin No. 302, Ames, Iowa, June 1942.

Hussain S.M., "A note on farmer's response to price in East Pakistan", *Pakistan Development Review*, Vol. 4, No. 1, Spring 1964.

Humphrey, Don, "The price of food", in A.M. Khusro (Ed.) *Readings in Agricultural Development*, Allied Publishers Private Ltd., New Delhi.

India, Government of, *Crop Weather Studies*, Bulletin No. 1, Ludhiana district (Punjab), Directorate of Economics and Statistics, Ministry of Food and Agriculture, New Delhi, 1970.

India, Government of, *Report of the National Commission on Agriculture, Part IV: Climate and Agriculture*, Ministry of Agriculture and Irrigation, New Delhi, 1976.

India, Meteorological Department, *Rainfall and Droughts in India*, IMD Report, 1971.

Iyengar, N.S., *Prediction of Food Production Including Rainfall Uncertainties*, Department of Science and Technology, Indian Institute of Sciences, Bangalore.

Jabar, M.A., "A note on the price response studies in relation to jute", *Indian Journal of Agricultural Economics*, Vol. 27, No. 1, January–March 1972.

Jacob, S.M., "Correlation of rainfall and the suceeding crops with special reference to the Punjab", *Memoirs of the Indian Meteorological Department*, Vol. 21, No. 1, 1916.

Jain, R.C., et al., "Effect of climatic variables on rice yield and its forecast", *Mausam*, Vol. 31, No. 4, 1980.

Jakhade, V.M. and N.A. Majumdar, "Response of agricultural producers to price—the case of jute and rice in India", *Indian Journal of Agricultural Economics*, Vol. 19, No. 3, July–December 1964.

Jha, Dayanatha, "Acreage response of sugarcane in Factory areas of North Bihar", *Indian Journal of Agricultural Economics* Vol. 25, No. 1, January–March 1970.

John, P.V., "Responsiveness of relative area and output of sugarcane and rice to changes in their relative prices in Uttar Pradesh: 1954–63", *Indian Journal of Agricultural Economics*, Vol. 20, No. 1, January–March 1965.

Johnson, D.G., *Agricultural Price Policy and International Trade*, International Finance Section, Department of Economics, Princeton University, June 1954.

Johnson, D.G., The nature of supply function for agricultural products", *American Economic Review*, Vol. 40, September 1950.

Johnson, G.L., "Supply functions: Some facts and notions", in Earl O. Heady et al., (Eds.) *Agricultural Adjustment Problems in a Growing Economy*, Iowa State College Press, Ames, 1959.

Johnston, J., *Econometric Methods*, McGraw Hill Book Co., New York, 1965.

Jones, G.T., "The response of the supply of agricultural products in the United Kingdom to price", *Farm Economist*, Vol. 10, No. 1, 1962.

Kahlon, A.S. et al., "Structure of farm prices in the Punjab", *Indian Journal of Agricultural Economies*, Vol. 20, No. 1, 1965.

Kainth, G.S., "Acreage Response of wheat in Punjab", *Journal of Social and Economic Studies*, Vol. 3, No. 3, July–September 1986.

Kainth, G.S., "Constraints to Agricultural Development", *Kisan World*, Vol. 21, No. 2, February 1994.

Kainth, G.S., *Foodgrain Marketing System in India: Its Structure and Performance*, Associated Publishing House, New Delhi, 1982.

Kainth, G.S., Foodgrains Prices in India: 1971–85", *Indian Economic Journal*, Vol. 35, No. 2, October–December 1987.

Kainth, G.S., "Growth of Indian agriculture: issues and policies", *IASSI Quarterly Journal*, Vol. 11, No. 2, October–December 1992.

Kainth, G.S., "Impact of state intervention in marketing", *Yojana*, Vol. 24, No. 7, April 16, 1980.

Kainth, G.S., "The impact of price movements on area under selected crops in Punjab", *Agricultural Marketing*, Vol. 20, No. 4, 1978.

Kainth, G.S. and A.S. Kang, "Price responsiveness of Indian Punjab wheat yield—a macro view", *Agricultural Situation in India*, Vol. 45, No. 12, March 1987.

Kainth, G.S. and R.S. Bawa, *Economic Development and Structural Changes: An Empirical Investigations*, Inter India Publications, New Delhi, 1985.

Kainth, G.S. and P.L. Mehra, *Rice Production: Potential and Constraints—A Case Study of Productivity Backward Region*, Inter India Publications, New Delhi, 1985.

Kakde, J.R., *Agricultural Climatology*, Metropolitan, New Delhi, 1985.

Kalamkar, R.J. and V. Satakopan, "The influence of rainfall distribution on the cotton yields at government experimental, farms at Akola and Jalagaon", *Indian Journal of Agricultural Sciences*, Vol. 10, 1940.

Kalyanasundaram, V. and K.S. Ramasastry, D*rought Index Based on Rainfall Deficiencies as Applied to Winter Rice in Bihar State,* Pre-published Report No. 97, Indian Meteorological Department, 1969.

Kamala, Devi P. and R. Rajgopalan, "Price and acreage response—a case study of groundnut crop in North Arcot district", *Indian Journal of Agricultural Economics*, Vol. 20, No. 1, January–March 1965.

Kamala, Devi P. and R. Rajgopalan, "Response of acreage to changes in price: a study in Madras State", *The Economic Weekly*, Vol. 16, No. 38, September 19, 1964.

Kanbur, M.V. and M. Neudekar, "Spatial equilibrium analysis, bufferstocks and support prices", *Artha Vijnana*, Vol. 10, No. 1, March 1968.

Kanithar, N.V., *Dry Farming in India,* Indian Council of Agricultural Research, New Delhi, 1960.

Kaul, J.L., "Acreage response to prices for major crops in Punjab—an econometric study", *Indian Journal of Agricultural Economics*, Vol. 26, No. 4, October–December 1971.

Kaul, J.L., "A study of supply response to price of Punjab crops", *Indian Journal of Economics*, Vol. 48, Part I, No. 188, July 1967.

Kawakatsu, S., "Crop production index number and their additive explanatory components", *Monthly Bulletin of Agricultural Economics and Statistics*, October 1969.

King, A.J., *The Crop Weather Problem—Research on Relationship of Weather to Crop Yields*, US Department of Agriculture, Bureau of Agricultural Economics, Washington, D.C., June 1938.

Klein, L.R., *An Introduction to Econometrics*, Prentice Hall, New Delhi, 1969.

Kothare, N.R., "Rainfall in India", *Reserve Bank of India Staff Occasional Papers*, Vol. II, No. 2, December 1977.

Kottke, M., "Discussion: the supply function in agriculture revisited", *American Journal of Agricultural Economics*, Vol. 51, May 1969.

Koyck, L.M., *Distributed Lags and Investment Analysis*, North Holland Pub. Co., Amsterdam, 1954.

Krishna, Jai, "Supply response and support prices—a model for determining support prices for agricultural commodities", *Agricultural Situation in India*, Vol. 22, No. 1C-11, January–February, 1968.

Krishna, Jai and M.S. Rao, " Dynamics of acreage allocation for wheat in Uttar Pradesh: a study in supply response", *Indian Journal of Agricultural Economics*, Vol. 22, No. 1, January–March 1967.

Krishna, Jai and M.S. Rao, "Price expectation and acreage response for wheat in Uttar Pradesh", *Indian Journal of Agricultural Economics*, Vol. 17, No. 17, January–March 1962.

Krishna, Raj, *Farm Supply Response in the Punjab (India-Pakistan)—A Case Study of Cotton*, unpublished Ph.D. dissertation, University of Chicago, 1961.

Krishna, Raj, "Farm supply response in India, Pakistan: a case study of the Punjab region", *Economic Journal*, Vol. 73, No. 291, September 1963.

Krishna, Raj, "Agricultural price policy and economic development", in Herman H. Southworth and Bruce Johnson (Eds.) *Agricultural Development and Economic Growth*, Cornell University Press, Ithaca, 1968.

Krishna, Raj, "Rice Zone policy: a note of dissent" in "*Report of the Agricultural Price Commission on Price Policy for Kharif Cereals for 1965/66 Season*, Ministry of Food and Agriculture, Government of India, New Delhi, May–July 1965.

Krishna, Raj and G.S. Raychaudhri, "Some aspects of wheat and rice policy in India", *World Bank Staff Working Paper,* No. 381, 1980.

Krishnaji, N., "Measuring agricultural growth", *Indian Journal of Agricultural Economics*, Vol. 35, No. 2, April–June 1980.

Krishnan, A. and B.V.R. Rao, "System analysis approach for crop planning—a case study of groundnut in Karnataka", paper presented at the symposium on E*ffects of Weather on Agricultural Production*, University of Kalyani, Kalyani, West Bengal, February 1979 (mimeo.)

Krishna, K.S. and C.L. Malik, "On statistical techniques for the study of crop, soil, weather and water relationship", *Journal of the Indian Society of Agricultural Statistics*, Vol. 22, No. 2, December 1970.

Krishnan, Thippalassery N., *The Role of Agriculture in Economic Development: An Econometric Study of Price Rela-*

tionships in India, 1950–62, unpublished Ph.D. dissertation, Massachusetts Institute of Technology, Cambridge, Mass, 1964.

Kumar, Pushpendra, "Income elasticity of demand for sugar—a regional analysis", *Artha VIjnana*, Vol. 9, No. 2, June 1967.

Lahiri, A.K. and P. Roy, "Rainfall and supply response", *Journal of Development Economics*, Vol. 18, 1985.

Lele, Uma J. and John W. Mellor, *Estimates of Change and Causes of Change in Foodgrains Production, India*, 1949–50 to 1960–61 Cornell International Agricultural Development Bulletin No. 2, 1964.

Lipton, Michael, "Should reasonable farmers respond to price dangers", *Modern Asian Studies*, Vol. 1, Part I, January 1967.

Lipton, Michael, "The theory of the optimizing peasant", *The Journal of Development Studies*, Vol. 4, No. 3, April 1968.

Lwarence, H. Shaw, "The effect of weather on agriculture output: a look at the methodology", *Journal of Farm Economics*, Vol. 46, 1964.

Madan, B.K., "Some observations on the essentials of a price policy of foodgrains in India, "*Indian Journal of Agricultural Economics*, Vol. 16, No. 4, October–December 1959

Madhavan, M.C., "Acreage response of Indian farmers: a case study of Tamil Nadu", *Indian Journal of Agricultural Economics,* Vol. 27, No. 1, January–March 1972.

Maity, S.P. et al., " The influence of climatic parameters on evapo-transpiration of crops", paper presented at the symposium on *Effects of Weather on Agricultural Production*, University of Kalyani, Kalyani, West Bengal, February 1979.

Maji, C.C. and D. Jha, "Cobb-Webb phenomenon to fluctuations in sugarcane acreage in North Bihar", *Indian Journal of Agricultural Economics*, Vol. 36, No. 4, October–December 1971.

Maji, C.C. et al., "Dynamic supply and demand models for better estimations and projections: an econometric study of major foodgrains in the Punjab region", *Indian Journal of Agricultural Economics*, Vol. 26, No. 1, January–March 1971.

Majumdar, A. and D.K. Bhattacharyji, "Effect of farmers decision and behaviour in substituting areas under jute and paddy in West Bengal", *Indian Journal of Agricultural Economics*, Vol. 20, No. 1, January–March 1965.

Mallik, A.K., "An examination of the crop yields at crop weather stations with special reference to rainfall", *Indian Journal of Meteorology and Geophysics*, Vol. 9, No. 1, July 1958.

Mallik, A.K., "Effects of weather on crop growth and yield at government experimental farm at Dharwar", *Journal of Biological Sciences*, Vol. 1, No. 1, 1958.

Mallik, A.K., "Height and yield of kharif jower in relation to rainfall during vegetative growth", *Indian Journal of Meteorology and Geophysics*, Vol. 9, No. 4, October 1958.

Mallik, A.K. and P.V. Pimpalwadkar, "Sugarcane crop in relation to weather", *Indian Journal of Sugarcane Research and Development*, Vol. 7, No. 2, January–March 1963.

Mallik, A.K. et al., "Preliminary studies on crop weather relations", *Indian Journal of Meteorology and Geophysics,* Vol. 11, No. 4, October 1960.

Mallik, R.P.S., *Influence of Rainfall and Temperature on Crop Production*, Agricultural Economics Research Centre, Research Study No. 80/4, Delhi, 1980.

Malya, Meenakshi M. and R. Rajagopalan, "Nature of risk associated with rainfall and its effect on farming—a case study of Kurnool district, Andhra Pradesh", *Indian Journal of Agricultural Economics*, Vol. 19, No. 1, January–March 1965.

Mann, Harold H., *Rainfall and Famine—A Study of Rainfall in The Bombay Deccan*, 1865–1938, The Indian Society of Agricultural Economics, Bombay, 1955.

Manning, H.L., "Calculation of confidence limits of monthly rainfall", *Journal of Agricultural Science*, Vol. 47, Part II, 1956.

Manning, H.L., "Confidence of limits of expected monthly rainfall", *Journal of Agricultural Science*, Vol. 40, 1950

Mason, E.S., Economic Planning in Underdeveloped Areas: Government and Business, Fordham University Press, New York, 1958.

Mason, E.S., *Economic Development in India and Pakistan,* Centre for International Affairs, Harvard University, Cambridge, 1966.

Mathur, M.B., *A Study of Movement of Prices of Selected Items of Foodgrains and Industrial Raw Materials in India*, 1939 to 1967–68, Occasional Paper No. 13, Cornell University—USAID, Employment and Income Distribution Research Project, Department of Agricultural Economics, Cornell University, 1968.

Mathur, P.N. and H. Ezekeil, "The marketable surplus of food and price fluctuations in a developing economy", *Kyklos*, Vol. 14, 1961.

Mavi, H.S., *Introduction to Agrometeorology*, Oxford and IBH Publishing Co., New Delhi, 1986.

McQuigg, James D. and John P. Doll, "Economics analysis involving random weather inputs", *Journal of Farm Economics*, Vol. 43, No. 4, November 1961.

Mellor, John W. et al., *Developing Rural India: Plan and Practice*, Cornell University Press, Ithaca, 1968.

Mellor, John W., *Agricultural Price Policy in the Context of Economic Development*, Occasional Paper No. 27, Cornell-USAID Prices Employment and Income Distribution Research Project, Department of Agricultural Economics, Cornell University, August 1969.

Mellor, John W., *Note on Agricultural Price Policy—1968 Indian Wheat Price Support*, Occasional Paper No. 5, Cornell—USAID Prices Employment and Income Distribution Research Project, Department of Agricultural Economics, Cornell University, January 1968.

Mellor, John W. and Ashok Dar, "*Changes in Relative Prices of Agricultural Commodities, India*, 1952–53 to 1964–65, Occasional Paper No. 1, Cornell—USAID Prices, Employment and Income Distribution Research Project, Department of Agricultural Economics, Cornell University, October 1967.

Mellor, John W. and Ashok Dar, "Determinants and development implications of foodgrains prices in India, 1949-50 to 1963-64", *The American Journal of Agricultural Economics*, Vol. 50, No. 4, November 1968.

Mellor. John W. and Ashok Dar, "Increasing agricultural production in early stages of economic development: relations, problems and prospects", *Indian Journal of Agricultural Economics*, Vol. 17, No. 2, April–June 1962.

Mellor, John W. and Ashok Dar, "The functions of agricultural prices in economic development", *Indian Journal of Agricultural Economics*, Vol. 22, No. 1, January–March 1968.

Mellor, John W. and Uma J. Lele, "Growth linkages of the new foodgrain technologies", *Indian Journal of Agricultural Economics*, Vol. 27, No. 1, January–March 1973

Mellor, John W. and Uma J. Lele, "Towards a theory of agricultural development", in Herman H. Southworth and Bruce T. Johnston (Eds.) *Agricultural Development and Economic Growth*, Cornell University Press, Thaca, 1968

Mellor, John W. and Uma J. Lele, "*Notes on Foodgrain Prices in India*, 1967-68 to 1968-69", Occasional Paper No. 2, Cornell USAID Prices, Employment and Income Distribution Research Project, Department of Agricultural Economics, Cornell University, October 1967.

Mellor, John W. and Uma J. Lele, "Alternative estimates of the trend in Indian foodgrains production during the first two plans", *Economic Development and Cultural Change*, Vol. 13, No. 2, January 1965.

Mellor, John W. and Uma J. Lele, *Agricultural Prices in Economic Development—Their Role Function and Operation*, Occasional Paper No. 39, Cornell—USAID Prices, Employment and Income Distribution Research Project, Department of Agricultural Economics, Cornell University, June 1970.

Mellor, John W. and Uma J. Lele, "The basis for agricultural price policy", *War on Hunger*, Vol. 4, No. 10, October 1970.

Minhas, B.S. and A. Vaidyanathan, "Growth of crop output in India: 1951–54 to 1958–61—an analysis by component elements, *Journal of Indian Society of Agricultural Statistics*, Vol. 17, No. 2, December 1965.

Minhas, B.S. and T.N. Srinivasan, "New agricultural strategy analysed", *Yojana*, Vol. 10, January 1966.

Minhas, B.S. et al., "Towards the structure of a production function for wheat yields with dated input of irrigation ratio", *Water Resources Research*, Vol. 10, No. 3, 1974.

Misra, V.N. and R. Radha Krishna, "Econometric models for acreage response: a study of groundnut and bajara in Gujarat", *Anvesak*, Vol. 1, No. 1, November 1971.

Morey, P.E., "Crop-yield-weather relationship studies in India by using multiple regression technique", paper presented at the 12th conference of the *Indian Econometric Society*, Kanpur, December 1972 (mimeo.)

Morey, P.E. and Madnani, M.L., *Regression Equations for Forecasting the Yield of Paddy on the Basis of Weather Parameters and Chemical Fertilizers—Tamil Nadu*, Meteorological Monograph, Indian Meteorological Department, 1973.

Morey, P.E. et al., *Forecasting Bajra with the Help of Weather Parameters*, Meteorological Department, Indian Meteorological Department, 1973.

Morey, P.E. et al., *Forecasting Yield of Principal Crops in India on the Basis of Weather Parameters—Paddy/Rice in North and South Assam*, Meteorological Monograph, Indian Meteorological Department, 1973.

Morey, P.E. et al., *Forecasting Bajra with the Help of Weather Parameters*, Meteorological Monograph, Indian Meteorological Department, 1973.

Mukherjee, Chandan and A. Vaidyanathan, "Growth and fluctuations in foodgrain yields per hectare—a statewise analysis", *Indian Journal of Agricultural Economics*, Vol. 35, No. 2, April–June 1980.

Mukhopadhyay, S.K., *Sources of Variations in Agricultural Productivity: A Cross Section Time Series Study in India*, The MacMillan Company of India, Ltd., New Delhi, 1976.

Mukhopadhay, S.K., *Effect of Weather on Agricultural Production: An Econometric Measure*, University of Kalyani, Kalyani, West Bengal, 1978 (mimeo.)

Myint, H., *The Economics of Developing Countries*, Hunchinson, London, 1964.

Nandkarni, M.V. and R.S. Deshpande, "Rainfall in Karnataka—some important issues in rainfall distribution", *Agricultural Situation in India*, Vol. 41, July 1982.

Nandkarni, M.V. and R.S. Deshpande, "Under-utilization of land—climatic or institutional factors", *Indian Journal of Agricultural Economics*, Vol. 34, No. 2, April–June 1979.

Nandkarni, M.V. and P.K. Ghosh, "Instability in rainfall and agricultural yields in a drought prone district (Tumkur)", *Indian Journal of Agricultural Economics*, Vol. 338, No. 2, April–June 1978.

Nair, K.R. and P. Bose, "Influence of humidity and temperature on the yields of cotton", *Sankhya*, Vol. 7, No. 2, November 1945.

Narain, D., *The Impact of Price Movements on Areas Under Selected Crops in India: 1900–39*, Cambridge University Press, Cambridge, 1965.

National Council of Applied Economic Research, *Long Term Projections of Demand and Supply of Selected Agricultural Commodities: 1960–61 to 1975–76*, New Delhi, 1962

National Council of Applied Economic Research, *Projections of Demand and Supply of Agricultural Commodities*, New Delhi, 1969.

Nerlove, Marc, *The Dynamics of Supply: Estimation of Farmer's Response to Price*, John Hopkins University Press, Baltimore, 1958.

Nerlove, Marc, "Distributed lags and estimation of long run supply and demand elasticities", *Journal of Farm Economics*, Vol. 40, No. 2, 1958.

Nerlove, Marc, "Estimation of elasticities of supply of selected agricultural commodities", *Journal of Farm Economics*, Vol. 38, No. 2, 1956.

Nerlove, marc, "Use of Durbin-Watson statistics in appropriate situation", *Econometrica*, Vol. 34, January 1966.

Nurkse, R. *Problems of Capital Formation in Under Developed Countries*, Oxford University Press, Bombay, 1964.

Nurkse, R., "Trade fluctuations and buffer policies of low income countries", *Kyklos*, Vol. 27, No. 2, April–June 1972.

Okada, T., "On the possibility of forecasting the approximate yield of rice crop for Northern Japan", *Journal of the Meteorological Society of Japan*, Vol. 36, November 1917.

Oni, S.A, "Estimating supply responses via producers panel approach", *Indian Journal of Agricultural Economics*, Vol. 27, April–June 1972.

Oury, Bernard, "Allowing for weather in crop production model building", *Journal of Farm Economics*, Vol. 47, No. 2, May 1965.

Oury, Bernard, *A Production Model for Wheat and Foodgrains in France*, North Holland Publishing Co., Amsterdam, 1966.

Panse, V.G., "Why crop cutting surveys: some further comments", *Indian Journal of Agricultural Economics*, Vol. 19, No. 2, April–June 1964.

Parikh, A., "An econometric model of the agricultural sector of Indian economy", *Journal of Policy Modelling*, Vol. 4, No. 3, 1981.

Parikh, A., "Farm supply response: a distributed lag analysis", *Bulliten of Oxford University*, Inst. Econ. and Stat. Vol. 33, No. 1, 1971.

Parikh, A., "Market responsiveness of peasant cultivators; some evidence from Prewar India", *Journal of Development Studies*, Vol. 8, No. 2, January 1972.

Parikh, Kirit S. and N.S.S. Narayan, *Estimates of Farm Supply Response and Acreage Allocation: A Case Study of Indian Agriculture*, International Institute for applied System Analysis, Research Report No. RR-81-1, Austria.

Penman, H.L., "Natural evaporation from open water, bare soil and grass", *Proceedings of Royal Society of London*, Series A, No. 193, July 1948.

Pesaran, M.H., "On the General problem of model selection", *Review of Economic Studies*, 1974.

Pesaran, M.H. and A.S. Deaton, "Testing of non nested non-linear regression model", *Econometrica*, 1978.

Philp, Cagan,"The monetary dynamics of hyperinflation" in Friedman Milton (ed.) *Studies in the Quantity Theory of Money*, University of Chicago Press, Chicago, 1956.

Pillai, P.P., "Impact of price movements on area, productivity and production of paddy in the Kerala state", *Agricultural Situation in India*, Vol. 24, No. 9, December 1969.

Pitter, R.L., "The effect of weather and technology on wheat yield in Oregon", *Agricultural Meteorology*, Vol. 18, 1977.

Prem Narain et al., "The use of Markov chain model for crop planning in rainfed areas", paper presented at the symposium on *Effects of Weather on Agricultural Production*,

University of Kalyani, Kalyani, West Bengal, February 1979.

Rajgopalan, V., *Supply Response for Irrigated Crops in Madras State, India*, Unpublished Ph.D. dissertation, University of Tennessee, 1967.

Ramamurthi, K.S. and J.R. Banerjee, "On the influence of weather on wheat yields at Dharwar", *Indian Journal of Meteorology and Geophysics*, Vol. 17, No. 4, October 1966.

Raman, C.R.V. and M. Srinivasa, *Water Availability Periods for Crop Planning*, Pre-published Scientific Report no. 173, Indian Meteorological Department, 1971.

Ramdas, L.A., "Fundamental facts of crop growth in relation to environment: precautions to be kept in mind in attempts to establish crop weather relationship by statistical treatment of data", *Journal of Indian Society of Agricultural Statistics*, Vol. 22, December 1970.

Ramdas, L.A., "Rainfall and agriculture", *Indian Journal of Meteorology and Geophysics*, Vol. 1, 1950.

Ramesh, D., "Long and short run elasticity of acreage under crops" *Agricultural Situation in India*, Vol. 20, No. 5, August 1965.

Ramesh, D., "A dynamic model analysis of foodgrains production and prices", *Agricultural Situation in India*, Vol. 19, No. 4, July 1964.

Rango Rao, D.S. and V.N. Pandit Rao, "A study of rainfall data at the four districts headquarter stations of former Bombay state", *Indian Journal of Agricultural Economics*, Vol. 14, 1959.

Rao, B.M., "A study of some weather factors on the yield of wheat in Ludhiana district, Punjab", *Indian Journal of Agricultural Economics*, Vol. 19, Nos. 3 & 4, July–December 1964.

Rao, G. Appa and B.S. Bhide, "Water availability periods and soil moisture during extreme years of crop production at Sholapur", *Mausam*, Vol. 31, No. 4, 1980.

Rao, G. Appa and S.N. Dudhane, "Weather features associated with abnormal wheat yields of Himachal Pradesh", *Mausam*, Vol. 35, No. 3, 1984.

Rao, G. Appa et al., *Weather Meters for Forecasting Wheat Yield in Madhya Maharashtra*, Meteorological Monograph, Indian Meteorological Department, 1973.

Rao, K.N. and J.C. Dass, *Weather and Crop Yields—Rice Survey*, Pre-published Scientific Report No. 137, Indian Meteorological Department, 1970.

Rao, K.N. et al., *Spectral Analysis of Drought Index (Palmer) for India*, Pre-published Scientific Report No. 169, Indian Meteorological Department, September 1971.

Rao, K.N. et al., "Nature of the frequency distribution of Indian annual rainfall and monsoon", Indian Journal of Meteorology and Geophysics, October 1972.

Rao, K.N. et al., *Indian Monsoon Correlations—Part I, Monthly inter correlations for all the meteorological sub divisions of India*, Meteorological Monograph No. 4, Indian Meteorological Department, Poona, 1972.

Rao, K.P.C. and V.K. Pandey, "Supply response of paddy in Andhra Pradesh", *Indian Journal of Agricultural Economics*, Vol. 31, No. 2 April–June 1976.

Rao, M.S. and Jai Krishna, "Price expectation and acreage response for wheat in Uttar Pradesh", *Indian Journal of Agricultural Economics,* Vol. 25, No. 4. October–December 1970.

Rao, V.M., "Farm production function studies: treatment of weather and selection of functional forms", *Economic and Political Weekly*, Vol. 5, No. 15, April 11, 1970.

Rao. V.M., "Methodological issues in measuring agricultural growth: lessons of recent researches", *Indian Journal of Agricultural Economics*, Vol. 35, No. 2, April–June 1980.

Ray, S.K., *Variations in Crop Output*, Institute of Economic Growth, Delhi, 1977 (mimeo.)

Ray, S.K., "Weather and reserve stocks for foodgrains", *Economic and Political Weekly*, Vol. 6, No. 239, September 25, 1977.

Ray, S.K., "Weather, prices and fluctuations in agricultural production", *Indian Economic Review*, Vol. 16, No. 4, 1981.

Rayner, A.J., "Positivistic measures of aggregate supply elasticities; some new approaches: comment", *American Journal of Agricultural Economics*, Vol. 52, February 1970.

Rayner, A.J., "The demand for inputs and the aggregate supply function for agriculture", *Journal of Agricultural Economics*, Vol. 21, May 1970.

Reddy, Mahender J., "Estimation of farmer's supply response a case study of groundnut", *Indian Journal of Agricultural Economics*, Vol. 25, No. 4, October–December 1970.

Reddy, V.N., "Growth rates", *Economic and Political Weekly*, Vol. 13, No. 19, May 13, 1978.

Roy, J. et al., *Handbook of Methods of Applied Statistics*, John Wiely and Sons, New York, 1967.

Roy, Shyamal, "Supply response of changes in price of jute in West Bengal," *Agricultural Situation in India*, Vol. 23, September 1968.

Rudra, A., "The rate of growth of Indian economy", in *Economic Development in South Asia*, Kandy conference Proceedings, edited by EAG Robinson and M. Kidron, Mac Millan, London, 1970.

Rudra, A., "Organisation of agriculture for rural development: the Indian case", *Cambridge Journal of Economics*, Vol. 2, No. 4, December 1978.

Ruttan, V.W. et al., (Eds.), *Agricultural Policy in an Affluent Society*, W.W. Norton & Company, Inc. New York, 1969.

Sanderson, Fred H., *Methods of Crop Forecasting*, Harvard University Press, Cambridge, Mass., 1954.

Sarkar, R.P., "A curvilinear study of yield with reference to weather-sugarcane", *Indian Journal of Meteorology and Geophysics*, Vol. 16, No. 1, January 1965.

Sarkar, R.P., "Crop weather relationship and crop yields forecasting", paper presented at the symposium on *Effects of Weather on Agricultural Production*, University of Kalyani, Kalyani, West Bengal, February 1979.

Sarkar, R.P. *Meteorology at the Service of Agriculture*, mimeo., Indian Meteorological Department, 1977.

Sarkar, R.P., et al., "Relation between growth and yield of sugarcane crop at Poona", Indian Journal of Meteorology and Geophysics, Vol. 13, No. 4, October, 1962.

Shah. V.C., "Agricultural price policy in a developing economy", *Indian Journal of Agricultural Economics*, Vol. 22, No. 3, July–September 1967.

Shenoy, B.R., "Foodgrains prices, controls and the plan", *Indian Journal of Agricultural Economics*, Vol. 12, No. 3, July–September 1957.

Shultz, T. W., *Transforming Traditional Agriculture*, Yale University Press, New Haven, 1954.

Shultz, T.W., *Economic Growth and Agriculture*, McGraw Hill, New York, 1968.

Shultz, T.W., *Economic Growth from Traditional Agriculture*, Publication No. 76, American Association for the Advancement of Science, Washington, D.C.

Shultz, T.W., *The Economic Organisation of Agriculture*, McGraw Hill, New York, 1953.

Shultz, T.W., "Reflections of agricultural production, output and supply", *Journal of Farm Economics*, Vol. 38, August 1956.

Sen, S.R., *Growth and Instability in Indian Agriculture*, Firma K.L. Mukhopadhyay, Calcutta, 1971.

Shah, C.H., "Comparison of yield estimates prepared on the basis of traditional and crop cutting methods", *Indian Journal of Agricultural Economics*, Vol. 17, No. 4, October–December 1962.

Shaha, S.K. and J.R. Banerjee, "Influence of rainfall, humidity, sunshine, maximum and minimum temperature on the yield of cotton at Coimbatore", *Indian Journal of Meteorology and Geophysics*, Vol. 26, No. 4, 1975.

Sharma, J.S., "Crop-weather relationship areas of study", *Journal of the Indian Society of Agricultural Statistics*, Vol. 22, No. 2 December 1970.

Sharma, P.S., "A study of variability of rainfall and cropping pattern in Gujarat", *Agricultural Situation in India*, Vol. 19, No. 8, November 1964.

Shaw, Lawrence H., "The effect of weather on agricultural output; a look at methodology", *Journal of Farm Economics*, Vol. 46, No. 1, February 1964.

Shaw, Lawrence H. and Durost D.D., *The Effect of Weather and Technology on Corn Yields in the Corn Belt: 1929–62, USDA*, Economic Research Service, Agricultural Economics Report 80, July 1965.

Shaw, Lawrence H. and Durost D.D., *Measuring the Effect of Weather on Agricultural Output: Procedure for Construct-*

ing Weather Indexes, USDA, Economic Research Service Report 72, October 1962.

Shourie, Arun, "The use of macro-economic regression models of developing countries for forecasts and policy prescription: some reflections on current practices", *Oxford Economic Papers* (New Series), Vol. 24, March 1972.

Sidhu, D.S. and J.L. Kaul, "Composite weather index for response studies", paper presented at the symposium on *Effects of Weather on Agricultural Production*, University of Kalyani, Kalyani, West Bengal, February 1979.

Singh, Avtar and M.V. Pavte, "Use of rainfall probabilities in agriculture and planning", *Indian Journal of Agricultural Sciences*, Vol. 38, No. 4, 1968.

Singh, Chokhey et al., "Yield of cotton in relation to rainfall in Madhya Pradesh", *Indian Journal of Agricultural Sciences*, Vol. 40, No. 1, January 1970.

Singh, I.J. and P. Kumar, "Impact of price and price variability on acreage allocation in Haryana", *Indian Journal of Agricultural Economics*, Vol. 31, No. 2, April–June 1976.

Singh R.D. and D.R. Rao, "Regional analysis of supply-price relationship of major foodgrains in Uttar Pradesh", *Indian Economic Journal*, Vol. 13, No. 4, October–December 1965.

Singh R.D. and D.R. Rao, "Determinants of supply behaviour–the case of virginia tobacco in India", *Artha Vijnana*, Vol. 16, No. 3, 1974.

Singh, R.D. et al., Estimation of agricultural acreage response relationship: some methodological issues", *Indian Journal of Agricultural Economics*, Vol. 29, No. 1, January–March 1974.

Singh, S.N. and S.L. Chowdhury, "Soil water in crop weather studies", paper presented at the symposium on *Effects of Weather on Agricultural Production*, University of Kalyani, Kalyani, West Bengal, February 1979.

Smith, B.B., "Relation between weather conditions and yield of cotton in Lousiana", *Journal of Agricultural Research*, Vol. 30, 1925.

Southworth, Herman M. and Bruce F. Johnston (Eds.), *Agricultural Development and Economic Growth*, Cornell University Press, Ithaca, 1967.

Sreenivasan, P.S., "Crop growth as the indicator of the final yield wheat crop", *Indian Journal of Meteorology and Geophysics*, Vol. 9, No. 2, 1958.

Sreenivasan, P.S., "Fisher's regression integral versus regression function of selected weather factors in crop weather analysis", *Indian Journal of Meteorology and Geophysics*, Vol. 23, No. 3, July 1972.

Sreenivasan, P.S., "The influence of rainfall on the yield of cotton in Madhya Pradesh", *Indian Journal of Agricultural Sciences*, Vol. 43, No. 9, September 1973.

Sreenivasan, P.S. and J.R. Banerjee, "A comparative study of rainfall and cotton crop under crop weather scheme at Akola and Nagpur", *Indian Journal of Meteorology and Geophysics*, Vol. 9, No. 2, April 1958.

Sreenivasan, P.S. and J.R. Banerjee, "A preliminary study of the growth and yield of sugarcane crop in relation to weather stations in Indian", *Proceedings of the Third Biennial Conference of Sugarcane Research and Development Workers* at Pusa, Bihar, 1957.

Sreenivasan, P.S. and J.R. Banerjee, "Studies on the forecasting of yield by curvilinear technique—rabi jower (sorghum) at Raichur", *Indian Journal of Meteorology and Geophysics*, Vol. 24, No. 1, January 1973.

Sreenivasan, P.S. et al., "Crop weather relationship studies", *Journal of Indian Society of Agricultural Statistics*, Vol. 22, No. 2, December 1970.

Srivastava, S.S., "Impact of rainfall on crop yield and acreage: a comment", *Indian Journal of Agricultural Economics*, Vol. 21, No. 2, April–June 1966.

Stallins, James L., "Weather and crop yield—a measure of influence of weather on crop production", *Journal of Farm Economics* Vol. 43, No. 5, December 1961.

Stallings, James L., "Weather indexes", *Journal of Farm Economics*, Vol. 42, No. 1, February 1960.

Stern, R.M., "The price responsiveness of primary producers", *Review of Economics and Statistics*, Vol. 44, No. 2, May 1962.

Subha Rao, K., "Farm supply response—a case study of sugarcane in Andhra Pradesh", *Indian Journal of Agricultural Economics*, Vol. 24, No. 1, January–March 1969.

Sud, Lalita and A.S. Kahlon, "Estimation of acreage response to price of selected crops in Punjab state", *Indian Journal of Agricultural Economics*, Vol. 24, No. 3, July–September 1969.

Sundaresan, R., *A Study on Variation in Rainfall and its Impact in Drought Prone Area of Tamil Nadu*, Research Report Ag. Econ./2/80, Centre for Agricultural and Rural Development Studies, Tamil Nadu Agricultural University, Coimbatore, 1980.

Srinivasan, T.N., "Constraints on growth and policy options: a comment", *Economic and Political Weekly*, Vol. 12, No. 48, November 26, 1977.

Srinivasan, T.N., "Trend in agriculture in India: 1949-50 to 1977-78", *Economic and Political Weekly*, Vol. 14, Nos. 30–32, August 1979.

Tippet, L.R.C., "On the effect of sunshine on wheat yield at Rothamstead", *Journal of Agricultural Sciences*, Vol. 16, Part 2, 1926.

Thornthwaite, C.W., "An approach toward a rational classification of climate", *Geographical Review*, Vol. 38, January 1948.

Tweeten, Luther and C. Leroy Quance, "Positivistic measure of aggregate supply elasticities", *American Economic Review*, Vol. 59, No. 2, May 1969.

Tyagi, D.S., *Farmer's Response to Agricultural Prices in India,* Heritage Publishers, Delhi. 1974

Vaidyanathan, A., "Constraints on growth and policy option", *Economic and Political Weekly*, Vol. 12, No. 38, September 17, 1977.

Vaidyanathan, A., "Constraints on growth and policy options: reply", *Economic and Political Weekly*, Vol. 12, no. 51, December 17, 1977

Vaidyanathan, A., "Influence of weather on crop yields: a review of agro-meteorologists research", *Indian Journal of Agricultural Economics*, Vol. 35, No. 2, April–June 1980.

Venkataraman, S., "Climatic considerations in cropping pattern", paper presented at the ICAR symposium on *Cropping Pattern in India*, Indian Council of Agricultural Research, New Delhi, 1968.

Venkataraman, S. and K. Narasimhamurthy, "On flowering behaviour of the rice crop at the crop weather stations in India", *Indian Journal of Meteorology and Geophysics*, Vol. 24, 1973.

Venkataraman, S.N., "Study of the effect of every inch of additional rainfall on lbs. per acre of cotton crop at Koilpatti", *Madras Agricultural Journal*, Vol. 18, 1930.

Venkataramanan, L.S., *A Statistical Study of Indian Jute Production and Marketing with Special Reference to Foreign Demand*, unpublished Ph.D. dissertation, University of Chicago, Chicago, June 1958.

Virmaki, S.M. et al., *Rainfall Probability Estimation for Selected Locations of Semi Arid India*, Research Report No. 1, ICRISAT, Hyderabad.

Wilcox, W.W., "Effect of farm price in efficiency in farming", *Journal of Farm Economics*, Vol. 33, 1951.

William, D.B., "Price expectations and reactions to uncertainty by illinois farmers", *Journal of Farm Economics*, Vol. 32, 1951.

Wolffram, W.F., "Price expectations and production plans", *Journal of Farm Economics*, Vol. 35, 1953.

Wolffram, Rudolf, "Positivistic measures of aggregate supply elasticities: some new approaches—some critical notes", *American Journal of Agricultural Economics*, Vol. 53, May 1971.

Waugh, Fredrick V., "Rainfall as an indicator of new England potato yield", *Journal of Farm Economics*, Vol. 11, No. 2, April 1929.

Wishart, J. and W.R. Macheniez, "Studies in crop variation VII: the influence of rainfall on the yield of Barley at Rothamstead", *Journal of Agricultural Science*, Vol. 20, 1930.

SUBJECT INDEX

AUTHOR INDEX